Mysterious Ocean

Peter Townsend Harris

Mysterious Ocean

Physical Processes and Geological Evolution

 Springer

Peter Townsend Harris
GRID-Arendal
Arendal, Norway

ISBN 978-3-030-15631-2   ISBN 978-3-030-15632-9   (eBook)
https://doi.org/10.1007/978-3-030-15632-9

Cover art work created by Eleri Mai Harris

This Springer imprint is published by the registered company Springer Nature Switzerland AG
The registered company address is: Gewerbestrasse 11, 6330 Cham, Switzerland

Prelude

Our perception of the ocean is surface deep.

The ocean is the blue horizon seen from the wave-battered coast. In our minds, the ocean surface *is* the ocean. We rarely think about what lies beneath the waves because the ocean is dark. It hides its secrets from our sight. The ocean is mysterious.

The moon, stars, and some planets are visible at night, while the ocean floor is invisible (out of sight, out of mind). Perhaps, this is one reason that the total investment in ocean exploration is only a fraction of the amount spent on exploring space. The reasons for this apparent bias are not economic. The offshore petroleum, shipping, and fisheries industries are clearly valuable to society (more valuable than space-based industries), so why is our investment in marine science not greater?

Is it because people believe we already know enough about the oceans, or as much as we need to know? If that is the case, then perhaps we need a reminder that we actually know very little about the oceans. We cannot explain with any certainty, for instance, how or when the oceans were formed. Geologists think it may have something to do with volcanic eruptions and comets hitting the Earth around 4 billion years ago, but they're not certain. Our map of the ocean floor is surprisingly crude; our images of the surface of Mars and Pluto are far clearer than what we have for most of the seafloor. We think the origins of the world's continental shelves (which produce 95% of all fish we eat and one-third of the petroleum we use) have something to do with rifting of supercontinents, sediment deposits, and changes in global sea level during the ice ages, but the puzzle is complicated and the story is not certain in most places. Tsunamis and earthquakes have caused incredible damage and loss of life. But although marine geologists developed the theory of plate tectonics around 1970 to explain why they occur, we cannot predict when they will happen with any certainty. We do not even possess an accurate map of the ocean's rift valleys, subduction zones, or plate boundaries, and we aren't exactly sure how the system works.

What about global climate change? The top 3 m of the ocean contain as much heat as the entire atmosphere, and the ocean contains 50 times more carbon dioxide than the atmosphere. Deep ocean circulation, a vital part of the Earth's climate system, is not that well understood by oceanographers. We think high levels of carbon dioxide in the atmosphere are being absorbed into the ocean making it more acidic, but we don't fully understand the transport of heat or storage of carbon dioxide and oxygen in deep ocean waters.

What about the diversity of life in the ocean? We think more species live in the sea than on land, but we don't actually know for sure. In 2010, marine biologists completed a global census of marine life and identified 230,000 marine species from about 30 million locations, but they also reported that there may be as many as 1 million species, mostly unknown to science. Ocean life is threatened by diverse human activities including global climate change, overfishing, pollution, and destruction of habitats. The cumulative impacts of these activities threaten the very existence of coral reefs, whales, and many important commercial fish species within the twenty-first century, but the full consequences are unknown.

In short, the oceans are a mystery to us. A brief survey of the unanswered questions that are currently being studied by marine scientists reveals the vast depths of our ignorance. In a world that is challenged by a growing human population needing more food, clean water, energy, and minerals while also needing to adapt to rising sea levels, climate change, degraded fisheries, earthquakes, and tsunamis, we cannot afford to remain ignorant about the oceans. Indeed, partly in recognition of this knowledge gap, the United Nations has proclaimed 2021–2030 the International Decade of Ocean Science for Sustainable Development.

This is a book about things that we don't know about the oceans but probably should. It is not a comprehensive review of ocean science by any means – that book would be far too long and is beyond my capability. I have instead tried to paint a broad picture of the oceans, their history, and the life they contain, with particular focus on unanswered questions based on my personal experience working as a marine geoscientist. There are some general facts about the Earth and ocean that I think everyone should know: For example, could you explain to a child how continents are formed? Do you know the name of the largest submarine canyon on the Earth? Do you know what the unit "Sverdrup" is a measure of? The most used commodity on the Earth is fresh water; do you know what the second most used commodity is? For answers to these questions, read on!

So why don't you join me on this tour of the ocean and its' 4-billion year history? Perhaps, it will make the ocean seem a little bit less mysterious. The ocean's story begins long ago, for there once was a time when the Earth did not have an ocean.

GRID-Arendal   Peter Townsend Harris
Arendal, Norway

Contents

Chapter 1
The Ocean Begins

*"Beginnings are apt to be shadowy, and so it is with the
beginnings of that great mother of life, the sea."*
Rachel Carson
The Sea Around Us, 1951.

Abstract The origins of the Earth and moon and the evidence for the first ocean
are discussed. We explore answers to the following questions: Were there any continents 4 billion years ago? What would the view from a beach look like 4 billion
years ago? How large were the first tides? Was the first ocean salty? What color
was the first sky? When did life first arise in the ocean? During its early history, the
Earth became a snowball; was the entire ocean ever frozen solid? What is the
"albedo effect"?

Keywords Solar system · Theia · Moon · Hadean · Ocean crust · Earth's core ·
Earth's atmosphere · Earth-moon distance · Ocean tides · Late heavy bombardment
· Banded iron formation · Photosynthesis · Great Oxygenation Event · Albedo ·
"Snowball" Earth · Huronian glaciation

In front of the building that houses *Geoscience Australia*, the national geoscience
agency in Canberra, there is an area of parkland that boasts a "geological time
walk." You can stroll along its 1100-m (3400 foot) length, where each footstep
represents about 3 million years of the Earth's four and a half billion-year history.
Shaded by eucalyptus trees, there are samples of rocks from the different geological ages, and signs are placed on the sides of the path noting various significant
events in the history of the Earth.

Close to the start of the walk, located about 4 billion years in the geologic past,
is a sign that reads: "The bombardment of the Earth by comets and meteors continued with such intensity that at times the oceans boiled." The thought of boiling
oceans, 4 billion years ago, seems quite amazing! But the other interesting point is

P. T. Harris, *Mysterious Ocean*, https://doi.org/10.1007/978-3-030-15632-9_1

that the oceans are truly ancient. Their existence dates back to the very earliest phase of the Earth's evolution.

The Earth is the only planet in our solar system that has an atmosphere and oceans of liquid water. Ganymede and Europa, two of Jupiter's moons, both have thin atmospheres, but the ocean on Europa is frozen at the surface. There is evidence that Ganymede has an ocean, but it is buried underground. Titan, the largest of Saturn's moons, has an atmosphere of nitrogen and seas on its surface, but they are probably made of liquid methane and ethane. Earth is unique in the solar system in having both an atmosphere and oceans made of water.

But it has not always been this way.

To tell the story of our oceans, we must start at the beginning of our planet's history, at the time when the oceans formed alongside the atmosphere. Earth's beginning is not completely understood, and there are many details of how the planet developed that are unknown. As Rachel Carson has pointed out, the ocean's beginnings are shadowy. There is of course very little hard evidence to go on, but we can work backward in time from what we know to have been the final products of the solar systems' formative processes to derive a working hypothesis of what is likely to have happened.

It is now believed that the Earth formed about four and a half billion years ago by accretion of gas, dust, and meteors at the time the solar system formed. The swirling mass arranged itself into the separate planets within the orbital plane, according to the laws of physics balancing the force of gravity against the centrifugal force. Matter was swept up by gravitational attraction, either by one of the orbiting planets or by the sun itself, which ignited once a critical mass was attained.

Around 100 million years after it formed, the Earth collided with a Mars-sized planet, romantically named "Theia" by scientists. The collision created a huge cloud of debris that coalesced and became Earth's moon. In Greek mythology, Theia is the mother of Selene, the goddess of the moon, so the name is appropriate since the collision of Theia and Earth gave birth to our moon.

The collision released so much heat that the Earth melted. This period is aptly known in geologic time as the Hadean Eon, which means "Hell like." It was a critical time in the formation of the Earth, because this is when the heaviest elements like iron and nickel sank into the interior forming a metallic core, and the different layers of molten mantle arranged themselves in order of increasing density. The lightest rock material then present, basalt, rose to the surface as erupting volcanoes and massive basalt flows. The surface of the moon retains basalt flows exposed on its cratered surface that have been dated to this approximate time (four and a half billion years ago), thanks to samples collected by the astronauts of NASA's Apollo missions.

As the Earth's surface cooled, the basalt hardened forming a crust. When geologists refer to the Earth's crust, they are talking about the solid layer of rock that is between about 5 and 40 km in thickness, which lies on the planet's surface. Beneath the crust we find the next layer (like the layers of an onion), called the mantle. The Earth is made mostly of mantle, about 84% by volume. The mantle is molten rock, and it becomes gradually hotter and less viscous the deeper into the Earth you go.

At the center of the Earth is a solid iron-nickel core. Its temperature is around 6000 °C. Heat from the center of the Earth is trapped, insulated by the upper layers of mantle and crust from the absolute zero temperature of space. The crust on the surface, exposed to the coldness of space (absolute zero is −273.15 °C!), forms as soon as enough heat is radiated back into space. But in the beginning of Earth's evolution, after the collision of Earth and Theia, there was no crust, and the mantle was exposed at the surface.

A solid layer of crust, a few kilometers thick covering the mantle, must have taken many millions of years to form and stabilize. This period of time is the only stage in Earth's history when there was no ocean.

When did the oceans form?

The secret of when the first ocean appeared on the Earth is revealed by sand grains comprised of the mineral zircon ($ZrSiO_4$). Zircon is a remarkably useful mineral. It is resistant to mechanical erosion and chemically insoluble in fluids, and its age can be radiometrically measured by the amounts of uranium and thorium that are incorporated into its structure at the time it crystallizes. The oldest zircon crystals that have ever been found on Earth are from a place called the Jack Hills in Australia, and they are 4.375 billion years old (give or take 6 million years).[1] This is the age when we first had igneous rocks (crust) on the surface of the Earth. The age of the oceans is revealed by the oxygen atoms inside the zircon crystal.

Oxygen comes as different isotopes: normal oxygen-16 (99.76% of atoms) and less common oxygen-18 (0.204% of atoms). The ratio of these two isotopes is very revealing. It turns out that zircons that are formed directly from mantle rocks have a very consistent oxygen isotope ratio. But zircons that formed in crustal rocks where water is present can have higher oxygen isotope ratios, greater than 30 per mil in some cases. Dating progressively younger zircons, we find that prior to 4.2 billion years ago, there are no zircons with oxygen isotope ratios greater than 5 or 6 per mil. But after 4.2 billion years ago, we find zircons having greater oxygen isotope ratios, indicating the presence of water. And the presence of water in the crust is explained by the presence of oceans. We can therefore conclude that there may have been oceans on the Earth since 4.2 billion years ago.

The first step in producing our oceans must have been to first produce an atmosphere, because without an atmosphere, there could be no liquid oceans. Water exposed to empty space would simply boil away because if there's not enough force (atmospheric pressure) to keep the water in a liquid phase, then there's nothing to hold the water molecules together. We can therefore deduce that the first water present was in the form of a gas (water vapor), mixed into the early Earth's atmosphere with other gasses. Earth's earliest atmosphere is thought to have been comprised of helium, hydrogen, and hydrogen compounds like ammonia and methane and probably some small amount of nitrogen.

But where did the atmosphere (and water) come from?

It is believed that the ocean and atmosphere originated from water molecules in the cloud of gas and dust that gave rise to the solar system. According to this theory,

---

[1] Valley et al. (2014).

as the Earth formed, water molecules became trapped in porous rock deep inside the molten planet. Then, for hundreds of millions of years, volcanoes erupted water vapor along with carbon dioxide, ammonia, and methane into Earth's atmosphere, enough to envelop the entire planet in a thick, gaseous blanket. Volcanoes are, to this day, still expelling water with each eruption, as measured by volcanologists. It is estimated that at present there may be enough water trapped in the mantle to fill the oceans three more times![2]

The surface of Earth gradually cooled over many millions of years until the crust solidified. Water vapor accumulated in the atmosphere, and when the surface temperature fell below 100 °C or 212 °F (at one atmosphere pressure), the first liquid water could be present on Earth. At just below this temperature, liquid water rained down onto the surface and pooled in depressions. This is when the oceans began to form. We can only speculate that scalding hot rain must have continued to fall for thousands of years, gradually filling the basins and eventually covering the entire surface of the Earth by around 4.2 billion years ago.

Water was also added to the atmosphere and oceans during the bombardment of Earth by comets and meteors. Meteors and comets are commonly composed partly of ice crystals that include frozen water, methane, and other compounds. Every comet impact added a little more water to the oceans as the basins gradually filled over millions of years. This is when the oceans boiled, around 4.2 billion years ago, as described in the geologic time walk outside the *Geoscience Australia* building.

At this early stage of the Earth's evolution, the basaltic crust was a relatively flat surface, devoid of the continental landmasses we have today. The oceans would have covered this surface to a more or less uniform depth. But to what depth? Or to put forward a simple question, was the volume of water greater or less than it is today?

Water is added to the oceans as it is expelled from volcanos and falls as rain. But water is also lost into space. When H_2O evaporates the separate hydrogen atoms and oxygen atoms are able to escape Earth's gravity. The high-energy Hadean sun could have caused a lot of water to have been lost from the early ocean.

Our knowledge of the Earth's early history is very fuzzy, but an intriguing idea is that, for some period of time, perhaps lasting several hundreds of million years, the oceans covered the entire surface of the planet. This idea stems from the likelihood that the Earth's continental crustal plates had not yet completely formed. And we also know that if the present land and seafloor surfaces were flattened to a single, globally uniform level, the volume of the oceans would cover the Earth to a depth of about 2600 m (5000 feet).

Astronomers have discovered that around one third of planets that are the size of Earth or larger have oceans; they are "ocean worlds." Earth's ocean accounts for about 0.02% of its mass (0.12% by volume), but some planets have oceans that are a much greater percentage. Imagine a world in which water accounts for 1% or 50% of the volume of the planet. Oceans on such planets would be hundreds or even thousands of kilometers deep!

---

[2] Schmandt et al. (2014).

Earth was fortunate to be able to keep its' ocean. Planets larger than Earth with stronger gravity lose less water into space, whereas smaller planets (like Mars) readily lose water and over billions of years are unable to retain their oceans. And so it is entirely possible that the early ocean on Earth may have been much deeper than it is today. Planet "Ocean" 4 billion years ago was the precursor to planet Earth.

Even when covered in ocean, Earth would not have been completely devoid of all land because there would have been a scattering of small continents and volcanic islands where the largest volcanoes rose from the seafloor to pierce the sea surface, like Hawaii does today. If you could stand on the shore of one of those islands, 4 billion years ago, the vista would bear little resemblance to our modern Earth.

***

What a scene you would behold standing on an ancient shore 4 billion years ago! The early atmosphere was rich in methane and hydrocarbon molecules, which would react to ultraviolet radiation to create an orange hue. Space probes have observed that orange is the color of the atmosphere on Titan, which is relatively methane-rich. Therefore, we can deduce that the ancient sky on Earth was probably orange.

The waves lapping the shore look reddish-pink in the sunlight. This is because the ancient ocean contained abundant dissolved iron. A decrease in dissolved iron will occur once life has evolved, but this has not happened yet.

The sun moves quickly across the orange-colored sky because the Earth revolved faster. A day lasted only about 16 hours – sunrise and sunset would only be 8 hours apart. The other extraordinary image would be the larger-sized moon, much larger than it appears today. This is because the moon was at least 20% closer after its formation and was perhaps only about half the distance that it is today.[3] The lunar surface facing Earth would be clearly visible at that close distance. On a clear night, the moon's craters and even erupting volcanoes may have been visible. Being closer to Earth, the moon orbited more quickly taking about 10 days at that time[4] compared with 29.53 days at present, so the phases of the moon would change rapidly, from new moon to full moon 5 days after that.

What caused the moon to move further away from the Earth, to the distance it is now? The answer is tidal friction.

The Earth rotates in the same direction as the moon orbits the Earth, but the Earth spins much faster than the moon orbits the Earth: 1 day versus one lunar month. As the Earth rotates, the face of one hemisphere moves toward the moon, while the other hemisphere is receding away from it. The gravitational pull of Earth is divided into the two hemispheres; the side moving toward the moon exerts slightly less gravitational pull, whereas the side that is receding is literally dragging the moon with it by the force of gravity (Fig. 1.1).

---

[3] It could not have been any closer to Earth than three times the Earth's radius, which is known as the *Roche limit*, the minimum distance from the center of the planet that a satellite can orbit without being destroyed by the severity of the tidal forces.

[4] Williams (2000).

Fig. 1.1 Diagram showing the rotation of Earth and lunar orbit viewed from above the North Pole. The red-dashed line represents the gravitational pull of the receding face of Earth which acts to pull the moon into a higher, longer-duration orbit

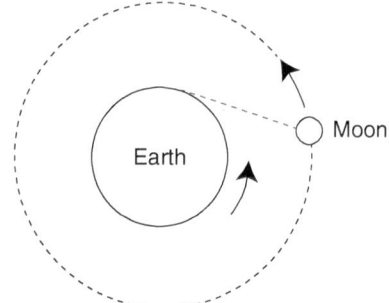

Over billions of years, this dragging force has caused the Earth's rotation to slow down so that 1 day now lasts 24 hours, while at the same time, it has pulled the moon into a faster, higher orbit around the Earth. With each orbit the moon has moved gradually further away. Using lasers, scientists have accurately measured the distance between the Earth and the moon over several years (the distance varies because of the moon's elliptical orbit). We know now that the moon is currently moving away from the Earth at a rate of 3.78 cm/year.

But there is a paradox – think about the collision of Earth and Theia. We know that right now the moon is (on average) 385,000 km from the Earth and computer modelling of that collision indicates that the moon afterward settled into an orbit that is closer to Earth. Since we also know how fast the moon is receding from Earth today, we can work backward to estimate the approximate timing of the collision of Earth and Thea.

And the answer is – around 1.5 billion years.

But the moon was supposed to have been formed over 4 billion years ago! This is the paradox. The logical explanation is that the rate of the moon in moving away must have been slower in the geologic past than it is today; how much slower, why, and when this happened are still further mysteries.[5]

Standing on that ancient shoreline 4 billion years ago, you would need to be careful how close you stood to the surf because the rising tide would be racing toward you at great speed. Ocean tides on the ancient Earth may have been extraordinary. The relationship between the Earth-moon distance and tidal height is nonlinear. At a distance of 200,000 km (i.e., about half the present distance), the ocean tidal range would have been more than 5 m (15 feet), and because of the shorter, 16-hour, day, the twice-daily tidal rise and fall was much faster. The rapid exposure and covering of intertidal flats are thought to have provided the perfect environment for the evolution of life.[6] When the Earth-moon distance was about 320,000 km,[7] the ocean tide would have been about twice its present (around 2 m) amplitude.

Would the sea water be salty or fresh?

The ocean 4 billion years ago was interacting with the thin ocean crust as shown by the oxygen isotopes stored in zircon crystals. This implies that crustal plates

---

[5] Bills and Ray (1999).

[6] Lathe (2004).

[7] Varga et al. (2006).

were subducting one over the other, bringing water into the mantle for recycling. Crust-forming rocks include chlorides, sulfates, and carbonates formed at low temperature (i.e., cooler than the mantle, within the crust), and leaching of these rocks supplies salt to the ocean. Perhaps there was a period of time when the oceans were freshwater followed by a later salty phase, but we don't really know for sure. Salt has been added gradually to the ocean from the time it formed; we'll come back to this topic later on.

The sky would provide an interesting spectacle, with lots of shooting stars. The so-called late heavy bombardment of the Earth and moon by meteorites was at a peak around 4 billion years ago and would not taper off for another 200 million years. Thanks to plate tectonics the Earth's surface is constantly renewed, and scars of impact craters do not linger. By contrast, the moon's cratered surface bears testament to impacts of millions of meteors from the late heavy bombardment and later times.

Standing on that ancient shoreline, the beach beneath your feet is made of black volcanic sand. There are no shells in the sand because mollusks and other shell-making organisms have not yet evolved, and there are no coral reefs or fish either. But there could be a strong breeze blowing waves on the ocean. Surfing would be feasible but probably a bit uncomfortable without protective clothing. That's because the ocean temperature might be a bit too hot for a pleasant swim; remember the oceans have only just recently been boiling from those comet impacts. Ultraviolet and solar X-ray radiation levels from the young sun were several orders of magnitude higher than today, and you'd have to hold your breath because the early atmosphere contained no oxygen. The best advice would be to wear a good spacesuit with built-in thermal regulator system, UV shield, and oxygen supply.

You might be standing on an island shore on the Earth, but the place you are standing is as alien and hostile to human life as the surface of Mars is today.

***

Hundreds of millions of years after the oceans boiled, life arose. There are dozens of books that speculate about how this may have happened. Amino acids, the building blocks of proteins, are easy to make. This was shown in the 1950s by the famous Miller-Urey experiment, which zapped a mixture of water and simple chemicals with electric pulses to simulate lightning strikes. More sophisticated experiments and theories have been developed since then, but there is no doubt that life did arise in the ocean because we have the evidence.

The oldest, indirect proof of the age of life in the oceans is recorded in 3.8-billion-year-old sedimentary rocks called the "banded iron formation" or "BIF" for short. These rocks consist of alternating bands of iron-rich (such as hematite and magnetite) and iron-poor (typically chert) sedimentary rock. The bands range in thickness from less than a millimeter to more than a meter.

Rocks of the banded iron formation are believed to have been deposited underwater, precipitating directly from seawater onto the seafloor. The precipitation of iron requires the presence of oxygen. And the source of the oxygen?

Life!

It is thought that the first oxygen came from microbes around 3.8 billion years ago that photosynthesized but that did not release free oxygen. The first free oxygen was produced probably before 3.5 billion years ago by photosynthesizing cyanobacteria, or blue-green algae, but the exact timing is not well understood. All such microbes and cyanobacteria must have been dwelling in the surface ocean because sunlight does not penetrate far into the depths, and all marine photosynthesizing organisms live in the "photic" zone, confined to the upper 100 m or so. The banded iron formation, therefore, tells us not only of the time of first oceanic oxygen but also the age of the earliest occurrence of photosynthetic microbes followed by early cyanobacteria.

The oldest banded iron formations are exposed in southwest Greenland at a location called *Isua*, and although they are also found at many other locations around the world, they appear to have only formed during certain phases of Earth's history. They first appear around 3.8 billion years ago, become common by about 3.5 billion, are abundant around 2.5 billion, and vanish by about 1.8 billion years ago. Banded iron formations made a brief comeback around 1 billion years ago, but none are being produced today.

In Earth's early history oxygen was absent from the atmosphere. Oxygen is a by-product of photosynthesis, so the Earth (and oceans) had to wait for the evolution of cyanobacteria (and eventually plants) before oxygen could be produced in sufficient quantities for the chemical process of oxidation to occur. The first living organisms neither produced nor consumed oxygen. Indeed, they would have been unable to tolerate the presence of oxygen: oxygen is a very reactive gas, and it is poisonous to organisms that are not adapted to its presence. For example, the bacterium *Clostridium botulinum* (a modern relative of early life) can only survive in the near-total absence of oxygen.

Life quickly adapted to make use of sunlight as a source of energy through photosynthesis that can be described by the chemical equation:

$$6CO_2 + 6H_2O \rightarrow 6O_2 + C_6H_{12}O_6$$

which in English means six molecules of carbon dioxide plus six molecules of water are transformed (by photosynthesis) into six molecules of oxygen (gas) and one molecule of carbohydrate (sugar). This simple formula underpins all life on Earth, and as we shall see, it has profound consequences for the evolution of our planet.

This leads us to an explanation for the iron-rich layers found in banded iron formations. Elemental iron dissolves in water, whereas the various oxides of iron (Fe_2O_3) precipitate out (become solids). The early oceans would certainly have had sources of iron, such as those emitted today from submarine volcanoes and liberated from rocks by chemical weathering (dissolving in water). The train of logic goes like this: when organisms arose that produced oxygen, iron that was dissolved in the oceans combined with dissolved oxygen to form iron oxides (the oceans "rusted") which would then have precipitated out and settled to the ocean floor. Blooms of planktonic cyanobacteria settled to the seafloor when they died, their decaying cells providing the silica (chert). Together these processes produced the layers of iron oxides and chert that characterize banded iron formations. The period of time when the banded iron formations were first deposited, about 3.8 billion years ago, is therefore the time when the oceans first contained dissolved oxygen.

Once most of the iron had been removed from the ocean, oxygen levels could build up until it was eventually added to the atmosphere. It was not until around two and a half billion years ago that the atmosphere contained its first trace amounts of oxygen, and it took another 2 billion years before the Earth's atmosphere changed from being oxygen-poor to oxygen-rich, like the atmosphere we have now.

Evidence suggests that atmospheric oxygen concentration was not more than about 10% up until around 800 hundred million years ago.[8] It is no coincidence that the Cambrian explosion in life on Earth did not occur until after this time, when oxygen levels rose to permit life that depends on breathing oxygen. Why did the atmosphere start to become oxygen-rich two and a half billion years ago? The simplest explanation is that it took around a billion years to use up all the free iron dissolved in seawater that was deposited in the banded iron formations. Once the iron was gone (or mostly gone), oxygen began to escape into the atmosphere. This episode in Earth's history, when oxygen first started to become rich in the atmosphere, is known by geologists as the *Great Oxygenation Event*. This event marked the end of the Archean eon and the beginning of the Proterozoic eon in the geologic timescale.

The Proterozoic is marked by the first appearance of "continental red beds" comprised of sediments deposited on land coated with iron oxide. These coatings must have formed during and/or immediately after their deposition which means that there had to be oxygen present in the atmosphere and in ground water.[9]

Let's put on our space suits again and go back to stand on that ancient shore and watch the ending of the Archean. The time is two and a half billion years ago. The view has changed quite a lot since the start of the Archean. There is a slight greenish tinge to the ocean today caused by all the plankton suspended in the water. The ocean tide has reduced to a more comfortable 3 m (10 feet) amplitude, so we can stand a bit closer to admire the surf, crashing on the beach. There is quartz sand on this beach made possible by the creation of granitic continents (more on that later) and a day now lasts 20 hours. Our young sun looks weak and pale compared with our sun today. That's because two and a half billion years ago, the sun's radiation was actually about 20% less intense than it is today. And there does seem to be a bit of a chill in the air; why is that?

The Earth's early atmosphere contained a lot of methane (CH_4) which is an incredibly powerful greenhouse gas, around 30 times stronger at trapping heat than carbon dioxide (CO_2). In the modern atmosphere, methane is quickly destroyed by free oxygen, so the methane level in our atmosphere today is very small, around one part per million (1 ppm). But since there was no free oxygen in the early atmosphere, methane was able to accumulate to very high levels, thereby warming the Earth and giving us liquid water and oceans while also turning the sky orange.

---

[8] Blamey et al. (2016).

[9] Walker (1979).

Recent chemical modelling of the Earth's early atmosphere suggests that methane levels were probably around 10 ppm.[10] By two and a half billion years ago, free oxygen was sucking methane out of the atmosphere at a tremendous rate. The reaction of oxygen and methane produces carbon dioxide and hydrogen. But since hydrogen gas (H_2) is light enough to escape Earth's gravity, only carbon dioxide, oxygen, and essentially inert nitrogen gas that dominate our atmosphere today were left behind (today's atmosphere is 78% nitrogen and 21% oxygen).

One visible sign of this change in atmospheric chemistry would be a familiar blue sky. But the oxygen levels are still too low for you to breathe, so you'll still need that space suit to survive on Earth.

On Earth two and a half billion years ago, the greenhouse gas levels (especially methane) began to rapidly decline, as the newly available free oxygen destroyed the methane gas. It is also possible that a lull in volcanic activity caused the levels of CO_2 to drastically reduce, further cooling the planet. At around 3 billion years ago, the atmosphere was probably only about half the thickness that it is at present,[11] and thinner air is colder (think of what happens when you climb a mountain to a higher altitude – where the atmospheric pressure is lower it gets colder!). But these are not the only factors causing a chill in the air. There is also the albedo effect.

The Earth's albedo (amount of sunlight that is reflected) is controlled to a large extent by how much ocean there is compared to the amount of land. Ocean has a very low albedo (absorbs heat from the sun) compared with a higher albedo on land (reflects more heat). Ocean near the equator absorbs the most heat, and land at the equator absorbs less (reflects more heat back into space). At times when the oceans covered more (or all) of the Earth, especially near the equator, then this would have a global warming effect. But when there is land along the equator, the Earth is cooled. The circumference of the Earth at the equator is 40,075 km (24,901 miles), and with the current (warm Earth) configuration of continents, 78.7% lies across water, and only 21.3% of the equator lies over land.

Another important factor controlling the Earth's albedo is cloud cover. White clouds reflect sunlight back into space cooling the planet by about 12 °C (22 °F). However, the water vapor comprising clouds is also a powerful greenhouse gas that traps heat and warms the planet by about 7 °C (13 °F). That is why cloudy nights are often warmer than clear nights. Think of the heat that comes off the pavement on a sunny day. After the sun goes down, you can still feel heat coming off the pavement – that's the longwave (invisible) radiation that clouds trap. The net result of cloud cover – cooling from albedo or warming from greenhouse – is a net cooling of about 5 °C (9 °F). The type of cloud (low versus high clouds and cloud density) also makes a big difference on both albedo and radiation trapping efficiency.

A combination of factors two and a half billion years ago favored a much cooler climate and the cooler climate allowed sea ice to form at the poles and ice sheets to cover the land at high latitudes. As the ice crept ever further south, year by year, the white albedo of snow and ice reflected the solar radiation back into space, cooling

---

[10] Olson et al. (2016).
[11] Som et al. (2016).

the Earth even more. And so, as the ice caps expanded, the Earth got colder and colder. The downward spiral toward an ever-colder climate led to a transformation of our planet.

Welcome to snowball Earth!

***

Once the polar ice caps expanded beyond a certain *tipping point* around 2.4 billion years ago, the Earth's climate became locked into a global ice age from which escape was (nearly) impossible. The concept of tipping points applies to economics, communications (when a tweet goes viral), the spread of infectious diseases, and climate and other natural cycles, including the population of species. Once a natural system has exceeded its tipping point, it can be very difficult to break the pattern and revert to a more balanced condition. It can also drive the system toward extreme conditions.

What extreme conditions occurred on snowball Earth? Did the oceans freeze throughout their depth with ice extending to the deepest parts of the seafloor? Were the oceans made of solid ice? How did Earth ever escape from this frigid situation?

We know from the rock record that glacial till was deposited on land located at tropical latitudes from between 2.4 and 2.1 billion years ago. This was the first global glaciation event, known as the *Huronian* glaciation, and it lasted for 300 million years.

Scientists now understand how finely balanced the Earth's climate actually is. A slight increase in solar radiation or greenhouse gas content in the atmosphere can tip the balance causing the climate to drastically change from greenhouse Earth to snowball Earth. There are many factors that exert control over the climate. Some climate change factors are very subtle like small changes in the Earth's orbit so that the sun is either closer or farther away. We have already discussed some others: changes in concentrations of trace gasses in the atmosphere, the Earth's albedo, and volcanic eruptions. Combinations of these factors can abruptly tip the climate balance from greenhouse Earth to snowball Earth, or vice versa.

How did Earth's climate recover from such extreme, snowball conditions? There may have been a different combination of the same factors, or perhaps other processes played a part. For example, we can assume that primary production must have slowed when the ocean was covered in ice. Sea ice blocks out the sun, and where the oceans were covered by year-round ice, primary production would be greatly reduced. In the seasonally melting sea ice zone at the poles, only specialized species of diatoms survive today. Since there was less primary production (there were no plants growing on land at this time), there was also less oxygen produced which, in turn, allowed greenhouse gasses like methane to slowly accumulate in the atmosphere once again.

Perhaps there was an episode of more intense volcanic activity, pumping more CO_2 into the atmosphere. And anyway, we don't really know how cold the Earth had become in the first place. The geologic evidence indicates that there may have been glaciers on land at tropical latitudes, just as there are glaciers today on the mountains of Papua New Guinea (in tropical latitudes). Perhaps "snowball" Earth looked more or less as it did at the peak of the last ice age, 20,000 years ago? We don't really know for certain.

Over the next 2 billion years, during the eon known as the Proterozoic, there is evidence that Earth experienced a global glaciation (snowball Earth) three or perhaps four more times. The exact causes of these glaciations, as well as the means of escape from snowball conditions, are unknown. However, we have a fairly good grasp on the basic principles that can drive the climate into snowball-Earth or greenhouse-Earth conditions. One factor we haven't discussed yet is the effect of continents on the oceans and on Earth's climate.

As noted already, continents have a higher albedo than the ocean, so when there is more land near the equator, there is more heat reflected and less heat absorbed by the oceans (and vice versa when there is less land and more ocean at the equator). Also, the continents play an important role in deflecting warm ocean currents toward the polar seas which can influence regional climates (like the Gulf Stream's effect on the climate of Norway). The continents wandered around the globe over the 300 million years of the Huronian glaciation, so it is possible that a continental configuration favorable to snowball Earth at 2.4 billion years changed to a less favorable configuration by 2.1 billion years ago. It is clear that we cannot tell the story of the oceans without also referring to the continents.

By the time of the Great Oxygen Event at the start of the Proterozoic, geologists estimate that Earth had accumulated about 40% of its continental area. Our planet was then 85% ocean and only 15% land. How was this land configured? Where did the continents come from in the first place? How did they grow bigger? How do we know that the continents have moved? The answer to these questions is provided by what is perhaps the most remarkable scientific discovery of the twentieth century.

The theory of plate tectonics.

Chapter 2
Oceans Created: Oceans Destroyed

"During the 1960s we found ourselves discarding most of our philosophy of the orderly development of the planet, and taking up what first seemed a prophetic dream of continents splitting apart and new oceans forming. Suddenly many puzzles of geological history began to make sense..."
Francis P. Shepard
Geological Oceanography, 1977.

Abstract Oceans have been created, and oceans have been destroyed many times in the Earth's history by plate tectonics. In this chapter we shall meet Alfred Wegner and learn about his early theory of continental drift and its problems. Rivals to Wegner's theory were the shrinking Earth theory and the expanding Earth theory, but they had problems of their own. We will meet Marie Tharp and her discovery of the great rift valley that encircles the globe. The "Rosetta Stone" for plate tectonic theory was the discovery of magnetic "stripes" on the ocean floor created by episodic reversals of the Earth's magnetic field. The disintegration of Pangea 170 million years ago and the rise and fall of the Tethys Ocean 6 million years ago are merely brief stages of the cycle of the birth and death of oceans.

Keywords Alfred Wegener · Pangea · Continental drift · Plate tectonics · Marie Tharp · Bruce Heezen · Harry Hess · Seafloor spreading · Paleomagnetism · Magnetic north pole · Panthalassa Ocean · Tethys Ocean

We are all standing on unsteady ground, for the continents are moving beneath our feet, driven by giant convection cells in the Earth's mantle. The continents float at the mercy of random upwellings and downwellings of molten rock, like the scum of milk floating on the surface of a cup of hot tea.

The discovery of plate tectonic theory involved many individuals who each contributed small pieces of knowledge that, once assembled, allowed the theory to develop over almost a century of scientific debate and exploration. No single person can take the credit for its discovery; there has never been a Nobel Prize awarded for the discovery of the plate tectonic theory. However, there have been a few individuals who have made exceptional contributions to developing the theory. One of them was Alfred Lothar Wegener (1880–1930).

Wegener was born in Berlin, Germany, and graduated from Friedrich Wilhelm University, completing his PhD in 1904 in the field of astronomy. Earlier workers had already noticed that the continents could be fitted together like pieces of a jigsaw puzzle; the west coast of Africa fits (almost) exactly into the east coast of South America, and the south coast of Australia fits into the coast of east Antarctica, etc. Geologists showed that, if the continents were assembled in a certain way, geological formations could be traced between them that contained fossils from the same geological era.

Wegener assembled vast amounts of data to develop and support his theory of "continental drift" which he later described in more detail in his book "The origins of continents and oceans" published in 1912. He did not stop there. He continued adding more and more information such that his book grew in size from 94 pages in its first (1912) edition to 234 pages in the fourth (1929); a sixth edition of the book was published in 1966 and it was translated into seven languages.

The meticulous gathering together of factual evidence that the continents had once been joined together in a giant supercontinent (named *Pangea* by Wegener) is undoubtedly one of Wegener's greatest contributions to the development of plate tectonic theory. That part of Pangea located at polar latitudes explained the occurrence of glacial deposits found today at sites near the equator; coal deposits that must have formed in humid tropical latitudes are today found in temperate and polar regions. Continental drift explained these observations. But Wegener's deductions went even further than this. He reasoned that mountain ranges must have formed where continents collided, and he gave the example of the Himalayas as an example (which is correct).

The influence of Wegener's theory was so large that two international conferences were organized, in 1923 (London) and 1926 (New York), although neither of the conferences was attended by Wegener himself. His PhD was in astronomy, not geology, a fact which detractors of the continental drift theory regularly pointed out. Wegner probably did not attend these conferences to avoid being ridiculed by his scientific rivals.

The prevailing, alternative theory, at the time Wegener published his continental drift theory, was that the Earth was shrinking and contracting due to cooling. Mountain ranges supposedly formed as the crust wrinkled, like the skin of a desiccated apple. The shrinking Earth concept had its own problems; for example, the shortening needed to raise all of the various mountain ranges required the volume of the Earth to have reduced by an amount much greater than could be explained by cooling and contraction of the crust and mantle. The shrinking Earth theory kept the continents fixed in their places, and this was the dogma of the scientific community well into the 1950s.

Wegener was convinced that the continents had once been joined in the geologic past but that they had "drifted" apart; the problem was that there was no logical explanation as to what force had caused the continents to "drift." Wegener never provided a plausible explanation. At first, he tried to explain drift in terms of tidal forces. In this model, continents are pulled by (lunar and solar) tidal forces such that they plowed through the ocean crust like ships through the ocean. Another suggestion was that centrifugal force caused the continents to drift toward the equator; subsequent changes in the Earth's axis of rotation accounted for the reason why not all continents have accumulated along the equator. But the physics did not make sense; tidal and centrifugal forces are far too weak to explain continental drift. These ideas as a driving force for the continental drift theory were rejected by the science community.

Wegener's scientific interests included meteorology and glaciology. He led four expeditions to Greenland to study the climate and measure the thickness of the ice sheet. During his fourth expedition to the Greenland ice sheet, Wegener led a team to resupply one of the bases. The weather closed in, the temperature dropped to −60 °C, and Wegener perished in a blizzard on September 24, 1930. He was 50 years old. His body was never recovered; presumably his remains are locked in the Greenland ice sheet and will one day emerge inside an iceberg. The Alfred Wegener Institute for Polar and Marine Research, founded in 1980 in Bremerhaven, Germany, honors the name of this pioneering earth scientist.

***

Wegener did not live to see his continental drift theory evolve into plate tectonics. But the fact that the continents were once joined into a giant supercontinent that Wegener had called Pangea was gradually accepted. A unifying theory that explained continental drift remained elusive. Continental drift was debated throughout the 1940s and 1950s when numerous ideas were proposed. But it was new scientific data collected from the oceans that finally made the difference. The first piece of critical information was a new generation of seafloor physiographic maps depicting the shape of the ocean floor.

This new way of viewing the ocean floor was invented by Marie Tharp and Bruce Heezen in their ocean panoramas. These pseudo, three-dimensional views of the ocean floor, based on available deep-ocean echo-sounding data, were published in the 1950s and 1960s culminating in a global ocean panorama published in 1977.[1]

Marie Tharp is one of the unsung heroes of the development of plate tectonic theory. Born in 1920 in Michigan, Marie was the daughter of Bertha, an instructor in German and Latin, and William, a soil geologist for the US Department of Agriculture. Marie took a job in 1948 at the Lamont-Doherty Earth Observatory in New York, working for the Director, Maurice (Doc) Ewing. Mapping the ocean floor was in its infancy, and Marie's job was to interpret data collected by the latest echo sounders used on research ships. Women were not allowed on research ships until the 1960s, so Marie worked on the data collected by others, including her collaborator, Bruce Heezen.

The basic approach was to transfer echo sounder profiles collected by ships onto a base map with a standard vertical exaggeration of 40 times. Where the echo-sounding lines were absent, Marie interpolated between them (used her

---

[1] Heezen and Tharp (1977).

imagination) to fill in the data gaps. Through her careful work, a new image of the ocean floor began to emerge. The ocean floor exhibits deep trenches close to the continents and thousands of volcanos strewn across the abyss. But it is clearly dominated by a mid-ocean ridge that snakes its way around the globe for 65,000 km (40,000 miles).

A critical observation was the occurrence of a central rift valley along the entire length of the mid-ocean ridge. Marie noticed that the rift valley coincided with the locations of seismic epicenters. The rift valley appeared on the physiographic map because Marie Tharp put it there. Others may take the credit for interpreting how the rift valley formed, but Marie Tharp discovered its existence and was the first person to map its course across the ocean floor.

By the end of 1956, the first physiographic map of the North Atlantic Ocean was completed. Bruce published papers in 1956 and 1957 that described the rift valley but failed to explain it. Neither paper included Marie as co-author. In 1957 Bruce was awarded his PhD, and he was invited to give a lecture on the physiographic map to faculty and students at Princeton University. He described the mid-ocean ridge encircling the globe with its continuous rift valley, the correlation of the rift valley with seismic epicenters, and the how it was linked to the African rift valley. At the end of the lecture, after a round of polite if not enthusiastic applause, the chairman of the geology department, Prof. Harry Hess, rose to his feet and said "Young man, you have shaken the foundation of geology!".[2]

***

Hess, Heezen, Ewing, and Tharp all recognized that the mid-ocean ridge with its central rift valley meant something – but what? How did it get there? What did Prof. Hess mean when he stated "Young man, you have shaken the foundation of geology!"?

One possible explanation for the apparent rifting of the ocean floor was that the Earth was expanding. The so-called expanding Earth hypothesis was promoted in the 1950s by Prof. S (Sam) Warren Carey of the University of Tasmania, in Australia. Bruce Heezen initially accepted the expanding Earth hypothesis as a logical explanation for the rift valley.[3] Problems with this hypothesis are that it requires the Earth's volume to be increasing exponentially in order to account for seafloor rifting, but there is no logical explanation for this supposed increase in volume. Furthermore, unless there has been a steady fall in global sea level, it requires that water must be added to the oceans at the exact same rate as the Earth expands in order for sea level to remain constant. But there is no evidence to suggest that global sea level has been falling exponentially in the Cenozoic or that current rates of water-sourced volcanism increased fast enough to fill the ocean basins at the required rate demanded by an expanding Earth.

The correct interpretation of the rift valley was first proposed by Harry Hess,[4] in his benchmark paper "History of ocean basins." In that paper Hess describes seafloor

---

[2] Felt (2012).

[3] Heezen and Tharp (1965).

[4] Hess (1962).

spreading, in which upwelling lava sourced from the mantle erupts along the rift valley. Hess recognized the significance of the rift valley which explains his out-burst at the end of Heezen's lecture in 1957. Seafloor spreading along the mid-ocean ridges pushes the continents apart as they ride passively over the mantle. Old ocean crust is subsumed beneath the continents where it descends and is subducted into the mantle. In Hess's words: "the cover of oceanic sediments and the volcanic sea-mounts also ride down into the jaw crusher of the descending limb."

Hess' seafloor spreading hypothesis also explained the observations that sedi-ments become thinner and younger in age, moving from land toward the mid-ocean ridge. The lack of thick sediment deposits in the ocean had perplexed geologists who had expected much thicker beds to occur there. They had (incorrectly) assumed that sediment had accumulated in the ocean from the time the Earth was formed. Subduction of ocean crust explained the small volume of sediments in the oceans compared with calculations of what should be present given measured rates of input from the land under uniformitarian conditions given the age of the Earth.

Hess described his work as "geo-poetry" because so much of what he had sur-mised was yet unproven. Still missing was any proof that the seafloor was actually spreading apart along the rift valley. It seemed possible or even likely that this was the right explanation, but direct proof was elusive.

The publication of "History of ocean basins" in 1962 marks the beginning of the establishment of plate tectonic theory, by providing a framework of working hypoth-eses backed up by existing data. A critical piece of that data is the physiographic map of the seafloor created by Bruce Heezen and Marie Tharp.

For people of my age, who were children in the 1960s, the mental image we have of the ocean floor is the one given to us by Marie Tharp and Bruce Heezen in their ocean panorama maps. Their maps came to us as blue-colored foldouts in copies of National Geographic magazine. We hung them on the walls of our bedrooms, and through Marie's vision, we imagined what the deep ocean floor really looked like. And we dreamed of one day going there ourselves.

In 1977 the global compilation of the ocean panorama was published, and in June of that year, Bruce Heezen died. He suffered a heart attack while on board a submarine exploring the rift valley just south of Iceland. He was just 53 years old. At the time of Bruce's death, Marie was aboard the British ship RRS *Discovery*, mapping the rift valley from above. She got the news by ship-to-ship radio. Marie never really got over Bruce's death. She continued drawing seafloor maps until she retired from Lamont in 1981. Marie Tharp passed away in August 2006.

During the years after 1960, more and more bathymetric data were gathered. Samples were collected and the composition of seafloor rocks and sediments were studied. But conclusive proof of seafloor spreading finally came from an unexpected source – the magnetic signature of the ocean floor.

***

The piece of evidence that can be said to have clinched plate tectonic theory came from studies of changes in the Earth's magnetic field, a discipline known as "paleomagnetism." The magnetic field of the Earth is generated by the motion of iron alloys in the liquid outer core. When lava erupts from volcanos on land or along the mid-ocean ridge, the iron-rich minerals align with the magnetic poles. Once the molten rock cools and solidifies, any iron particles it contains retain the magnetic orientation that prevailed at the time of cooling.

Two pioneers in the study of the Earth's magnetic field were the French geologist Antoine Joseph Bernard Brunhes (1867–1910) and the Japanese geophysicist Motonori Matuyama (1884–1958). Brunhes had noted that the orientation of the magnetic field preserved in some rocks is different from the present-day orientation. Either the landscape preserving these rocks had moved (changed orientation) or else the magnetic field had changed direction (or both). By recording different orientations of the magnetic field preserved in volcanic rocks of different ages, taken in the same fixed location, Brunhes demonstrated in 1905 that the magnetic poles must have switched orientation at some time in the past.

Meanwhile, Motonori Matuyama collected samples of basalt in Manchuria and in Japan and measured their magnetic polarity. The basalt samples collected from different sediment layers had different polarities which Matuyama deduced (in a paper he published in 1929) was caused by a reversal in the Earth's magnetic pole sometime during the Pleistocene (last 2 million years).

Later researchers have pinpointed the time of the first reversal to around 781,000 years ago, an event now known as the Brunhes-Matuyama reversal. In fact, we now know that the magnetic field has switched back and forth over geologic time, about once every 200,000 years or so on average, with magnetic north located for some period of time in the northern hemisphere and then in the southern hemisphere (magnetic north is presently in the southern hemisphere, at a point in the Southern Ocean off the coast of George V Land, East Antarctica).

The development of different instruments to measure the magnetic properties of rocks (called magnetometers) led eventually to a device that can be towed behind research ships during the late 1950s. These instruments measure the residual magnetic orientation of the ocean crust. Systematic oceanographic surveys towing magnetometers over mid-ocean spreading ridges produced the first maps of the seafloor's magnetic signature. The maps revealed that the basalt erupting along the rift valley axis preserves the modern magnetic alignment.

Since the magnetic field maintained its present orientation for the last 780,000 years, a period known as the Brunhes chronozone, the basalt that has erupted along the spreading ridge, has the modern (present-day) orientation. At some distance further away from the rift valley, the ship crosses the Brunhes-Matuyama reversal, and the magnetic field is aligned in the opposite direction. Towing a magnetometer back and forth across a spreading ridge reveals a symmetrical pattern of magnetic reversals on either side of the central rift valley. The data piled up until the evidence pointed to a remarkable conclusion.

The pattern is caused by seafloor spreading!

Magnetic data collected in the North Atlantic and in the Indian Ocean, followed by additional evidence from spreading ridges in the northeast Pacific off Canada's Vancouver Island, all revealed a symmetrical pattern of normal and reversed magnetism. Picture two conveyor belts with the ends facing each other so that the upper surfaces are moving apart (Fig. 2.1). The gap between the belts represents Marie

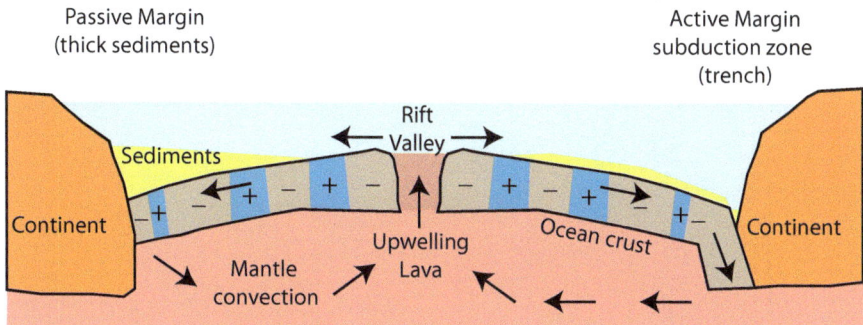

Fig. 2.1 Sketch showing seafloor spreading of ocean crust caused by lava upwelling from the mantle into the rift valley. Continents and ocean crust float on the mantle. Where ocean crust abuts the continents, it is either subducted beneath or else rests passively against continental margins. The blue and brown stripes represent positive (+) and negative (−) magnetic polarity of the solid ocean crust. Sediments are thickest adjacent to passive margins and thin or absent on the mid-ocean ridge

Tharp's rift valleys. The belts rise up and move apart, just as lava rises to the surface and forces the rift valley sides to spread. The lava cools and solidifies and the process repeats itself. Once there is a magnetic reversal, the erupting lava will magnetize with the opposite polarity. As the conveyor belts spread apart, the magnetic signature forms a mirror image of positive and negative magnetic orientations on each side of the rift valley.

The hypothesis that the so-called magnetic stripes are the result of seafloor spreading was proposed independently in 1963 by Frederick Vine and Drummond Matthews[5] in the United Kingdom and separately by Canadian Lawrence Morley. This not only proves the point that great minds think alike. It was a huge "Eureka!" moment for marine geoscience.

The idea that the magnetic stripes are proof of seafloor spreading (now called the Vine-Mathews-Morley hypothesis) was so controversial in 1963 that Lawrence Morley's paper describing the hypothesis was rejected by two scientific journals: *Nature* and the *Journal of Geophysical Research*. The geological community at the time contained many skeptics who doubted all these crazy newfangled ideas: seafloor spreading, geomagnetic reversals, and continental drift. These ideas seem, in hindsight, very obvious to us today, but in 1963 they were revolutionary.

Magnetic stripes are much more than direct evidence of seafloor spreading. They are like a "Rosetta Stone" for marine geologists. The symmetrical pattern of thick and thin magnetic stripes provides us with a measure of the duration of magnetic chronozones. At a constant spreading rate, wider stripes represent a longer chronozone than narrow stripes. Radiometric dating of the rocks and overlying sediments, taken together with the width of magnetic stripes, provides an estimate of the rate of seafloor spreading and how it has varied over time. Following their path

---

[5] Vine and Matthews (1963).

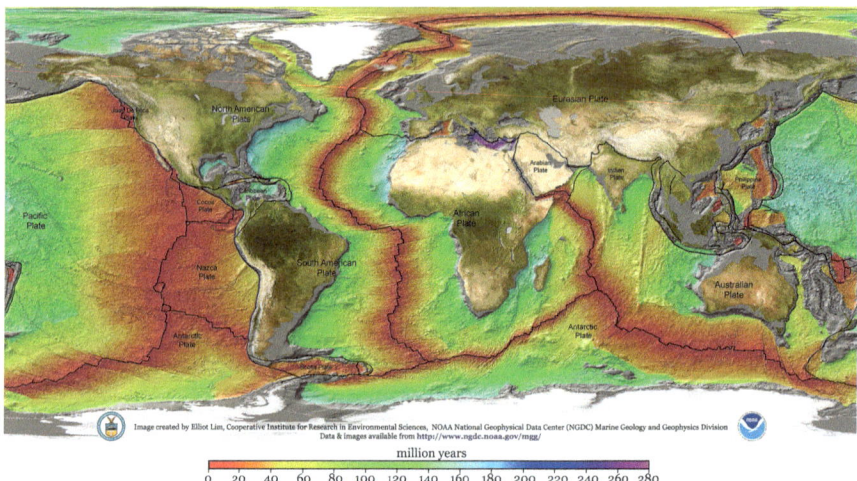

million years

0   20   40   60   80   100   120   140   160   180   200   220   240   260   280

Fig. 2.2 Age of ocean crust. (Map produced by NOAA)

away from the spreading ridge shows us the track over which the continents have moved, as well as their speed and direction. From the age of the oldest "stripe" adjacent to a continent, magnetic stripes tell us in some cases when a particular ocean basin began to form.

Suddenly, with the new insight provided by magnetic stripes, many parts of the plate tectonic theory fall into place. The stripes provide an accurate tool for tracking the movements of continents as they are pushed apart by spreading ridges. The creation of ocean crust along spreading ridges is balanced by its destruction when it is subducted back into the mantle, manifest on the seafloor as deep ocean trenches. The pressures exerted on the crust by the push/pull forces of spreading and subduction cause it to fracture, explaining the elongate faults that appear as ridges and valleys that are aligned normal to the axis of spreading.

The magnetic stripes also provide an accurate tool for measuring the age of ocean crust. From this we have learned that the ocean crust is relatively young, about 80 million years old on average (Fig. 2.2). It has this young age because the oldest ocean floor is destroyed when it is subducted back into the mantle. The oldest parts of the continents are over 4 billion years old, but the oldest ocean crust that currently exists, located in the Ionian Sea and the east Mediterranean, is only about 270 million years old.[6] The oldest ocean crust in the Pacific and Atlantic Oceans is around 180 million years in age. Plate tectonics explains how the water filling the ocean basins is around 4 billion years old, but the basins themselves are everywhere less than 270 million years old and mostly less than 180 million years.

In the pre-plate tectonic era of geologic thinking, the age of seafloor mountain ranges was a subject arousing much speculation and debate. Mountains that form on

---

[6] Muller et al. (2008).

land are subject to erosion by rain, snow, and ice. After only a few million years, the jagged peaks become rounded, reduced over geologic time to sand and mud that is transported to the ocean by rivers. But under the ocean, there is no rain or snow. The processes that erode mountains on land are absent under the ocean. And so it was believed that submarine mountains must be truly ancient, their origins dating back to the formation of the Earth itself. Plate tectonic theory explains the youth of submarine mountains that are composed of oceanic crust and that have lifespans limited to the time between their formation on the mid-ocean ridge and subduction in deep ocean trenches.

***

Using the tracks of magnetic stripes, Prof. Dietmar Muller and his colleagues at the University of Sydney in Australia have pieced together a sort of "time-lapse movie" of how the continents have migrated across the globe over the past 400 million years.[7] Based on the available evidence, we know now that Alfred Wegener was right; there was once a giant supercontinent called "Pangea" which existed when all the continents were squashed together into a single, large blob. At this time, around 350 million years ago, there was only one superocean, the Panthalassa Ocean, and one massive continent.

Pangea split apart around 250 million years ago and new oceans began to form. The first of these oceans, called "Tethys," lay between Laurasia (made up of North America and Eurasia) and Gondwana (South America, Africa, Antarctica, Australia, and India). During the early Cretaceous period, 130 million years ago, the Tethys Ocean extended between Laurasia and Gondwana in a continuous belt around the globe, between the equator and 30° north latitude.

North and South America began to part from Laurasia and Gondwana about 180 million years ago to eventually create the Atlantic Ocean. The supercontinent, Gondwanaland, was located over the South Pole, and this split apart around 100 million years ago into separate continents of Africa, Antarctica, Australia, and India. Antarctica stayed at the South Pole, but the other continents broke away and moved north. India broke off and shot across the Indian Ocean at a great speed of up to 36 cm/year before colliding with Asia causing the creation of the Himalayas (just as Alfred Wegener had predicted). The collision has slowed India down, and it is now only moving northward at around 2 cm/year. Australia moved northward at a more leisurely pace of 7 cm/year, and it is presently crashing into the Pacific Plate, building mountains on the island of Papua New Guinea. Africa moved northeast at a relatively sluggish speed of around 2 cm/year and is presently crashing into Europe, forming the Alps, and squashing the Mediterranean Sea.

What happened to the Tethys Ocean? It was destroyed by plate motions of India and Africa. The present-day Mediterranean Sea, the Black Sea, the Caspian Sea, and Aral Sea are the last vestiges of Tethys. The once great Tethys Ocean has been replaced by different oceans as the continents closed in and squeezed it out of existence. Tethys seashells can now be found on the peaks of the Himalayas.

---

[7] Muller et al. (1997).

Revealing the complex puzzle of how the continents have split apart and moved around the globe over the past 400 million years has taken decades of research and study. But 400 million years, vast as this length of time may seem, accounts for only about 10% of the history of the oceans. Without the detailed information on continental movements that ocean crust reveals through its magnetic stripes, geologists have only the rocks that remain on the continents to try to piece together the chronology of supercontinent assembly and disintegration and (from their paleomagnetic signatures) an approximation of their latitude and orientation. The names of these ancient supercontinents sound exotic and quite romantic: "Rodinia" formed about 1 billion year ago, "Nuna" about 1.8 billion, "Kenorland" 2.7 billion, and "Ur" formed 3 billion years ago.[8] The ocean crust that carried these continents across the face of the earth has long since been absorbed back into the mantle; we will never know the paths followed by these ancient continents as they wandered across the globe.

Each disintegration of a supercontinent saw the creation of new oceans, a new configuration of ocean basins, and each assembly of supercontinents saw the separate oceans merge again into a single, continuous world ocean. This great cycle of ocean creation and destruction through plate tectonics is called the "Wilson cycle," named for person who first described it, Tuzo Wilson.[9] There have been perhaps 20 or more Wilson cycles of oceans created and erased by the restless wandering of the continents over the last 4 billion years. What did those oceans look like? Their secrets remain a mystery that will probably never be solved.

The restless wandering of the continents across the face of the Earth has created and destroyed numerous oceans. And yet there has only ever been one ocean. The continents separate and divide the ocean into what looks like separate ocean(s), but it is all connected. It has always been, in a sense, the same ocean.

Over geologic time the area of land comprising the continents has grown larger which means that on a planet of finite size, the area of the oceans must be decreasing. But how can the continents be growing larger? Where did they come from in the first place?

---

[8] A movie showing the wandering continents over 3 billion years can be viewed here: https://www.youtube.com/watch?v=ovT90wYrVk4

[9] Wilson (1966).

Chapter 3
Where Did Continents Come From?

"Science, my lad, is made up of mistakes, but they are mistakes which it is useful to make, because they lead little by little to the truth."
Jules Verne
A Journey to the Center of the Earth, 1864

Abstract Did you ever wonder where the continents came from? Their creation is related to plate tectonic subduction zones and the melting of ocean crust. The subduction zones are like a giant "refinery" which results in the gradual accretion of continents. In this chapter we will learn how to think in "deep time": thousands, millions, and billions of years. The first sexual reproduction occurred in the ocean less than a billion years ago, but for over 3 billion years before that, the oceans contained strange-shaped mounds made of sediment called "stromatolites" created by cyanobacteria. The Cambrian Period saw the explosion of life, but the fossil record we have is very patchy because the chances for any plant or animal to become a fossil are less than one in a million. The formation of the modern ocean basins began only around 180 million years ago – isn't it strange that the ocean floor and ocean basins are so much younger than the ocean water they contain?

Keywords Subduction zone · Stratovolcano · Pillow lava · Zealandia · Deep time · Stromatolites · Sexual reproduction · Cambrian explosion · Census of Marine Life · Fossil · Sedimentary rock · Trenches · Spreading ridges · Wilson cycle

High school science education provides most people with an understanding of certain key scientific concepts that we should all understand. Darwin's theory of the evolution of life, the "Big Bang" theory of the origin of the universe, how the light bulb was invented, etc. are useful and important ideas that we should all know and be able to explain to our children. But it is a remarkable fact that the average person has no idea how Earth's continents evolved. How can you not know about the origin of the very ground you are standing on?! Surely the evolution of our planet's oceans and continents is just as important to understand as the evolution of life on our planet?

To understand how the Earth got its continents, we must simply apply the theory of plate tectonics. The story begins with ocean crust. The solid basalt rock that forms ocean crust is a very thin veneer, only around 5–10 km thick. It overlies the molten rock (magma) comprising the mantle. Remember, the temperature of Earth's core is around 6000 °C, compared with the temperature of oceanic crust (which is close to 0 °C on the ocean floor) so there is a huge vertical gradient in temperature. The hot magma rises, transporting heat upward toward the surface, but the magma is trapped below the solid ocean crust. In places where the ocean crust is thinnest, it is lifted up and splits apart, creating mid-ocean ridges with their distinctive rift valleys.

Along the rift valley, the seafloor fractures and splits apart, partly due to the force of gravity as the crust slides down the flanks of the ridge and partly due to the pressure of upwelling magma. Immediately after fracturing, magma rises to quickly fill the cracks, and it cools and solidifies, over and over, again and again. The magma cools and crystalizes adding more ocean crust, and the process continues.

As the ocean crust is transported further and further away from the spreading ridge, it cools, contracts, becomes denser, and subsides. Old ocean crust is cooler, denser, and located in deeper water than younger ocean crust. The process is slow. The ocean crust moves only a few centimeters per year, and the spreading ridges are many 10's of kilometers wide.

Seafloor spreading is the first driving process of plate tectonics. Subduction is the second process. As the seafloor spreads apart, horizontal pressure builds up within the ocean crust until it buckles, with one layer overriding another. Small differences in the density of the crust may play a part in determining which layer rides over the top of the other. The bottom layer sinks down into the mantle, dragging bits of the upper layer with it, forming a deep trench in the seafloor. The role of seawater is critical in the process – water provides a lubricant that eases the path of the subducted plate beneath the overlying plate. It is not clear that subduction of ocean crust is even possible without water lubricant (does this mean that planets without oceans don't have subduction zones?). The great ocean trenches are places where ocean crust is being subducted back into the mantle.

The pulling force of the crust that is sinking into a trench (subduction zone) is the second driving force of plate tectonics. Geologists are uncertain which process is the dominant one: the seafloor spreading "push" along the mid-ocean ridges or the "pull" caused by crust sinking into subduction zones of the upper mantle. It is not a coincidence that the Atlantic Ocean, which has few deep trenches and subduction zones (e.g., the Puerto Rico Trench) and has mostly passive margins, has an average rate of seafloor spreading of around 15–20 mm/year, whereas the Pacific Ocean, which is surrounded by trenches and subduction zones and active margins, has an average seafloor spreading rate of 40–60 mm/year. Both processes, spreading and subduction, are important and fundamental to the evolution of continents and oceans according to plate tectonic theory.

Now we come to the part of the story that explains why we have continents on the Earth. The sinking ocean crust is made of basalt that contains a number of minerals, including iron and silicon. As the crust descends into the subduction zone, it gradually gets hotter and hotter. But the temperature at which the different minerals melt varies. Silicate rocks melt first as the crust sinks, with iron-rich minerals remaining solid. Since silicon is less dense than basalt, it rises upward upon melting, while the remaining crust, depleted of silicon, sinks into the mantle. Silica (quartz) and aluminum are the major components of granitic rocks that form continental crust.

Pressure builds up during subduction of the crust, and volcanos erupt, bringing the silica-rich minerals upward to the surface. Although the exact process that moves the granites upward from the mantle into the huge blocks forming the continents is not completely understood, in essence the deep ocean trenches and their subduction zones are like giant refineries. In these refineries, less dense granitic rocks are distilled away from the melting ocean crust to form continental land masses, while the denser, basalt-rich crust sinks to be melted and returned to the mantle. The subduction refinery liberates other elements like nitrogen that do not fit easily into the crystal lattice structure of minerals, and so it escapes from volcanos into the atmosphere as nitrogen gas. Very stable and unreactive, N_2 (nitrogen gas) is also a heavy molecule (it does not escape easily into space from Earth's gravity like hydrogen does), and so it has gradually accumulated in the atmosphere over geologic time.

Our atmosphere (78% nitrogen) is a by-product of plate tectonics.[1]

The volcanos that form on continental margins are very different from the ones that erupt on the ocean floor. So-called stratovolcanos (also called composite volcanos) are built up over time by layers of hardened lava, tephra, pumice, and volcanic ash containing high amounts of silica. Their eruptions are far more violent than underwater basaltic volcanos because silica is stickier and more viscous than basalt when it melts. Sticky lava makes a kind of volcanoes, near subduction zones, that are explosive when they erupt (think of Mount St Helens and Krakatoa). The purely basaltic volcanos that erupt on the ocean floor are quiet and passive by comparison. Their molten basalt oozes out slowly forming "pillow lavas" (because they are shaped like a pillow), which have been filmed as they have formed underwater in Hawaii.

Over billions of years, the continuous subduction of ocean crust leads to more and more distillation of the granitic rocks that comprise the continents. It is not a one-way process; subduction of oceanic crust may occasionally drag chunks of the continents down with them into the mantle. Continental abrasion may have exceeded accretion in some locations over some lengths of time. There may have been some small continents present when Earth was formed, or none at all. The details are all a bit fuzzy. But what we do know is that over billions of years of continuous seafloor spreading and ocean crust subduction, the continents have accreted, bit by bit, from the partial melting of basaltic seafloor as it is subducted back into the mantle.

---

[1] Mikhail and Sverjensky (2014).

The continents are a by-product of plate tectonics.

Much of the early history of the continents is uncertain.[2] Since the Earth's crust was probably thin at first (in the Hadean Eon), it is not clear if there were any crustal plates. Perhaps some continental crust was produced quickly, close to the surface during the Hadean. If there were plates, they were probably thinner and moved more quickly than they do today since the mantle convection system probably moved faster in the Hadean. Did the continents evolve rapidly at first and then more slowly as the processes of plate tectonics became established and also slowed down? Or did the evolution of continents progress steadily, incrementally, growing piece by piece over the eons?

You may recall from Chap. 1 that the oldest zircon crystals (from the Jack Hills of Australia) are around 4.4 billion years old. This date is just 100 million years after the collision event with Theia melted the Earth's surface. Therefore, the formation of granite occurred almost immediately after the formation of the first oceanic crust. If granite requires the existence of subduction zone refineries for its formation, it seems logical to conclude that the evolution of spreading ridges and subduction zones must have occurred at around this same time.

The first continents, therefore, emerged at about the same time as the oceans were formed. If the first continents were formed along plate boundaries as theory suggests, then their shape would be elongate, forming stringy-shaped micro-continents along the edges of their adjacent ocean trenches. New Zealand or Japan might be good examples of what small, long, stringy continents looked like.

If a continent is too small, it might not even rise above sea level. The height of a continent above sea level, its "freeboard," is proportional to its area. The larger a continent is the higher its freeboard.[3] The mean elevation of Asia (the largest continent) is 600 m (2000 feet), whereas the mean elevation of Australia (the smallest continent) is 250 m (820 feet). There are several factors that interact to give this result: the collision of continents and subduction of ocean crust make the continents thicker so they ride higher; erosion by rivers and glaciers erodes taller mountains faster than lower ones; larger continents trap more heat rising from the mantle below, and so they float higher on average than smaller-sized continents (Fig. 3.1).

If the volume of water filling the oceans was comparable to the amount on Earth today, then the earliest, mainly small-sized continents were probably mostly submerged. Examples exist today of submarine plateaus studded with island archipelagos. In fact, there are several broad submarine plateaus attached to the New Zealand "micro-continent," the whole of which is known as "Zealandia."

As more and more material was added from the subduction refinery, the first micro-continents emerged from the oceans and grew larger. Continents are transported across the globe, floating over the hot molten mantle, riding passively on their ocean-crust conveyor belts. From time to time, two or more micro-continents collided and merged forming a larger continent. At other times, continents found themselves located over a mantle upwelling zone and were summarily rifted apart.

---

[2] Nagel et al. (2012).

[3] Zhang (2005).

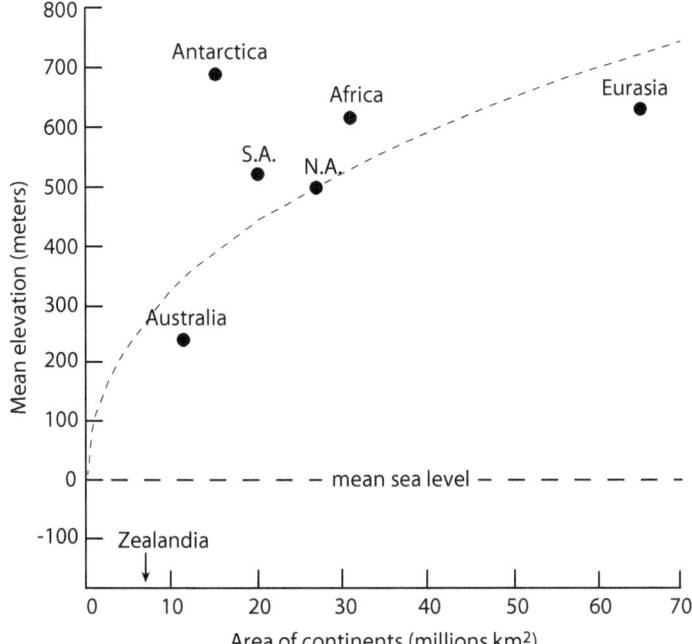

Fig. 3.1 The mean elevation of continents versus their area (redrawn from Zhang 2005, to include Zealandia which lies at a mean depth of about 1000 m below sea level), showing how the elevation (or depth of submergence) of continents and continental fragments is proportional to their area

Over geologic time, the continents accreted more and more granite from the subduction refinery. By dating the age of granitic rocks, we can determine that around 40% of the continents had formed by the end of the Archean (before around two and a half billion years ago), another 40% by the end of the Proterozoic (by around half a billion years ago), and the last 20% or so formed during the Phanerozoic (from half a billion years ago to the present).[4]

The continents merged with other continents and mini-continents to make supercontinents (when most of the land area was lumped together into a single land mass) or else they split apart making many smaller-sized continents. The continents we see today have a 4 billion-year history of repeated cycles of accretion, merging and rifting. The thickness of continental crust is typically 30–40 km (compared with 5–10 km for ocean crust), attesting to the long history of the accumulation of the continental masses.

***

Thinking in terms of hundreds of millions or billions of years is what geologists refer to as "deep time" (Fig. 3.2). In order to grasp the concept of geologic (deep)

---

[4] Kearey et al. (2009).

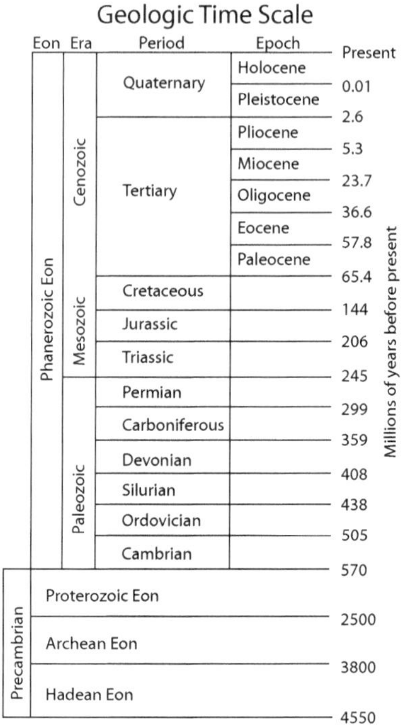

Fig. 3.2 Simplified geologic timescale (from USGS). Note how most of the geologic Periods date from less than 570 million years ago – most of what we know about Earth's history is from the Phanerozoic Eon, and we have much less knowledge of the Precambrian

Geologic Time Scale

Eon	Era	Period	Epoch	Millions of years before present
Phanerozoic Eon	Cenozoic	Quaternary	Holocene	Present
				0.01
			Pleistocene	
				2.6
		Tertiary	Pliocene	5.3
			Miocene	23.7
			Oligocene	36.6
			Eocene	57.8
			Paleocene	65.4
	Mesozoic	Cretaceous		144
		Jurassic		206
		Triassic		245
	Paleozoic	Permian		299
		Carboniferous		359
		Devonian		408
		Silurian		438
		Ordovician		505
		Cambrian		570
Precambrian		Proterozoic Eon		2500
		Archean Eon		3800
		Hadean Eon		4550

time, let's try a thought experiment. We shall travel back in time on a logarithmic scale (1, 10, 100, 1000 years, etc.).

Think of the street where you live. Can you describe what has changed on your street over the past year? Probably, you can easily recall several details of what may have changed. Perhaps a house was painted, or a new building went up on a vacant lot, or a new street sign was put up.

What about 10 years ago? Have there been any significant changes on your street in the last decade? Most people could probably recall a few things, perhaps houses were sold, neighbors put up a new fence, a corner grocer went broke, and maybe the road was re-paved. Ten years is within our personal, living memory, so you can rely on your own firsthand observations to recall what things were like on this timescale.

To go back in time, 100 years in time is beyond living memory (at least it is for most people!). Grandparents or great grandparents might recall a few facts about the neighborhood. You might try and find old photos of your street, perhaps in the local library. Many of us live in suburbs that didn't exist 100 years ago. The land where your street is may have been used for farming or some other purpose at that time.

To go back in time, 1000 years is most likely before your street existed unless you are lucky enough to live in a very old town. This timeframe is getting into the realm of archeology, and to find out what your street looked like 1000 years ago would require some serious research. Perhaps your street is in a town that has been occupied by several different civilizations over time. In places where settlers came from different continents, the trees and shrubs seen today may be totally different from 1000 years ago. For most people, the street where we now live was in the wilderness 1000 years ago.

To go back in time 10,000 years, we leave the timeframe of civilization and its archeology. A geologist might be able to tell you something about the land where your street is located by studying the fossils and pollen preserved in local sedimentary strata. Perhaps your street was in an area that had been covered by ice during the last glaciation; imagine what the landscape looked like at the time just after the ice had retreated. Perhaps 10,000 years ago, your street was in a forest prowled by large predators or in open grassland grazed by bison. The sedimentary record might be able to tell you these things, but there will be nothing in any history books.

To begin to think of what your street looked like 100,000 years ago, from a human perspective, is revealed by the research of physical anthropologists. Modern humans had only evolved in Africa around 200,000 years ago, so unless your street is in Africa, any *Homo sapiens* walking along what is now your street were among those early migrants who are believed to have walked to China and Australia by around 100,000 years ago. The environmental history of your street might be preserved in local sedimentary strata.

Can you imagine what your street looked like 1 million years ago? Or 10, or 100 million years ago? Thinking in these different, longer timeframes is what geologists term "deep time." Depending on the timescale you choose, some processes are very rapid or very slow. A human life span seems long thinking in terms of days or months, but very short in terms of decades. Human evolution is very slow at a timeframe of 100 years but rapid at a timescale of 100,000 years. The formation of the northern polar ice cap that corresponds with the ice ages takes place on timescales of 10,000–100,000 years. Plate tectonics and the relative movement of continental plates appear very slow at timeframes of 100,000 years, but rapid at timescales of tens to hundreds of million years.

Geologists think of different processes operating within different frames of "deep time," from the convection of the mantle, to plate tectonics and the growth of continents, to the evolution of life, the creation of a river estuary, and the cycles of the ice ages. When a geologist looks at a mountain range, for example, among her/his first thoughts will be their stage of formation and age. Young mountains with their tall, spikey peaks may be only a million or less years in age, whereas low, rounded peaks hint at a greater age, perhaps many 10's of millions of years following the end of their last mountain-building tectonic phase. Deep time is another way to think about the Earth and its history; it's the way we must think in order to grasp the history of the Earth and of the oceans.

The 1100-m-long geologic time walk outside the Geoscience Australia building (referred to in Chapter 1) makes this point very well. At the very end of the walk, there is an elongate sign about 4 m long, with some geological events marked out. Each meter is equal to about 4 million years, so humans only appear in the last few centimeters. There is no room to display anything about recorded human history at this scale, crammed as it is into the last 1 or 2 mm.

In the vastness of time that spans the history of the oceans, our personal experience of change is remarkably limited. The processes that have formed the oceans take place mostly on very long timescales, and the changes that we have lived to see firsthand are very few indeed. The problem is that we tend to distrust things that we cannot observe firsthand, which is one reason why people resisted accepting plate tectonic theory for so long. To think in terms of "deep time," we have to put aside our distrust of slow-moving processes and open our minds to the possibilities of what can occur given enough time. Mountains rise up and are eroded away, continents drift across the globe, ice sheets and glaciers expand and then disappear, life evolves, and the climate changes.

As an example of thinking in deep time, consider the physical processes that cause global sea level to rise and fall. Let's start with geologically slow processes and work our way toward faster-paced processes. The formation of continents occurs on timescales of 100s of millions to billions of years – bigger supercontinents like Pangea trap heat and float higher over the mantle such that sea level appears lower relative to the land.

Mid-ocean ridges rise up when heated by convection in the mantle displacing ocean water onto the land. The mid-ocean ridges have a total length of around 65,000 km (40,400 miles), but when ocean ridges covered more of the ocean floor (under a different configuration of continents and ocean basins), they may have raised sea level by 100 m (300 feet) or more. Conversely, when ocean ridges covered less ocean floor, sea level would have been lower.

In contrast with these slow, plate tectonic processes, the formation of continental ice sheets can occur on timescales of thousands of years, causing sea level changes of around 100 m (300 feet) in amplitude. Melting or creating larger alpine glaciers can occur over century timescales, causing sea level changes of around 1 m (3 feet). Warming of the oceans in the last century has caused the ocean to expand, raising global sea level by around 10 cm (4 inches) since 1950.

This leads us to one final aspect that distinguishes the geological sciences from other disciplines. The normal scientific process starts with a theory from which predictions are made that can be tested, usually in a controlled laboratory setting. This is how chemistry and physics experiments are done, for example. In the case of geology, however, many hypotheses relate to some aspect of the Earth that can only be "tested" by making observations in nature. The problem is that many geological processes are very slow-moving compared against the length of human life spans, let alone the amount of time a research grant will last (usually about 3 years). For example, the theory of plate tectonics predicts that the continents are moving in relation to spreading ridges, but moving very slowly. There was no way to measure these movements before the invention of the Global Positioning System (GPS), but now we have measured it. The two halves of Iceland, divided by the mid-Atlantic rift valley, are splitting apart at a rate of 2.5 cm/year based on 15 years of continuous GPS measurements.[5]

---

[5] Geirsson et al. (2010).

Theories about events that occurred in deep time are difficult to test. The rock record is very fragmented and discontinuous. We see snippets of time recorded in the layers of rock, and the scientific method is turned on its head – the rocks record something that happened in the past, and the geologist must essentially ask: "what was the experiment?"!!

The history of the ocean is recorded in rocks and sediments deposited over geologic time. We know that sediments deposited in the oceans record at most only the last 200 million years or so, because older sediments have already been subducted into the mantle. That leaves only the sediments and rocks on land as a resource to consult for Earth's history prior to 200 million years ago. We must rely on the random chunks of ancient seabed that were scraped off the ocean crust and plastered onto the continents to search for clues.

The evolution of life before 200 million years ago is a story written in the rock record found on the land. Evidence that life began in the oceans 3.8 billion years ago is recorded in the banded iron formations already discussed. How did those simple cyanobacteria cells evolve after that?

***

Evolution is a slow process. For the first 2 billion years of Earth's history (the Archean), there was no life in the oceans more complex than single-celled bacteria and amoebae. At the beginning of the Proterozoic, from around 2.5 to 2.1 billion years, was the "snowball" Earth, associated with the "Great Oxygenation Event" that we already discussed. Once the oxygen feast was over, however, it is now thought that O_2 was depleted from the ocean and atmosphere and it fell to very low levels, perhaps to less than 1% of the atmosphere. The low concentration of oxygen in the atmosphere did not support a protective ozone layer, and consequently UV radiation levels were very high. Life evolved slowly over the next billion years of the Proterozoic with remarkably little to show for the length of time involved.

The Earth entered an incredibly stable period of climate and tectonics. Two supercontinents called Columbia and Rodinia coalesced and then disintegrated. Mountains were lifted up and then eroded away by glaciers and rivers. The ocean currents flowed ceaselessly around the globe. The climate remained stable, and there were no further ice ages. For 1 billion years, from about 1.8 to 0.8 billion years ago, things on Earth and in the ocean were a bit *ho hum*. This time period is known by geologists as the "boring billion" or the "dullest time on Earth." The description seems a bit harsh, but there you have it.

But it was not completely boring during the boring billion. One or two events are worth mentioning. For example, the first sex took place in the ocean.

Around 1.2 billion years ago, the eukaryotes evolved containing a cell with a nucleus enclosed in a membrane (cyanobacteria cells do not have a nucleus). This nucleated cell structure enabled sexual reproduction by meiosis to occur for the first time. The first sex between two consenting cells was a big event during the boring billion.

The other highlights of the Proterozoic are the stromatolites. Cyanobacteria colonies, with some help from the branch of eukaryotes that include algae, created a kind of sedimentary rock that is the iconic symbol of the Archean and Proterozoic Eons. Stromatolites are essentially a biologically mediated sediment deposit, in which photosynthesizing cyanobacteria play a key role. The typical stromatolite is shaped like a column or mushroom, and it grows to a few feet in height in shallow intertidal waters. It grows vertically upward from the seabed when sediment grains become suspended in the water column (such as happens during a storm event). The suspended sediments become trapped within the bacterial matt. The cyanobacteria grow over the sediment, striving toward the sunlight, and thus form a growth layer on the upper surface. The next event adds more sediment followed by more bacterial growth, one layer on top of another.

Stromatolites first appeared in the Archean 3.85 billion years ago and were prolific in the Proterozoic, reaching a peak in abundance around 1.2 billion years ago. They only became rare when animals evolved that grazed upon them for the algae and bacteria growing on their surfaces. Today stromatolites are found in only a few places, like Shark Bay in Western Australia, where the extreme salinity of the water deters grazing metazoans that would otherwise consume the algal coating.

***

Standing on an ancient shoreline 1.2 billion years ago, the ocean looks a bit more like our modern world. A day now lasts around 21 hours. The beach is made of quartz sand, sourced from the erosion of continental mountains, and delivered by rivers flowing into the sea. There are no shells on the beach, but we can see stromatolites in the shallow, clear water. The water is clear because the oceans are not very productive; the water is not clouded by floating plankton. A bluish-green slime grows in the shade of rocks (in the shade to avoid harsh UV sunlight) along the coast, and the low hills in the distance also have a faint greenish tinge in places, but there are no plants or trees and no insects or animals of any kind.

The air has only a trace of oxygen and you would be unable to breath without an oxygen mask. Just as well since there is a strong smell of rotten eggs in the air. Hydrogen sulfide gas is escaping from the ocean due to the rapid influx of sulfate liberated from the oxidation of pyrite on land which peaked during the Great Oxygenation Event. Now oxygen levels in the ocean and atmosphere have dropped again; the boring billion corresponds to a gap in the formation of banded iron formation, pointing at the low levels of dissolved oxygen possibly combined with the oceans' depleted sources of iron. The Earth seems a rather barren, empty, and smelly world by comparison with ours.

The oceans formed 4.2 billion years ago, and it has taken 3 billion years to arrive at our current point in the midst of the Proterozoic "boring billion." Life evolved early on to make use of sunlight to create chemical energy through photosynthesis (the cyanobacteria) by around 3.8 billion years ago, and then it took another 2.6 billion years before a cell with a nucleus evolved.

Complex life like we see today did not appear on the Earth until around 600 million years ago. To put 600 million years ago into perspective, you will have completed 950 m of the 110 m geological time walk in front of the Geoscience Australia building to cover that span of Earth's history. That is to say, 87% of the 4.6 billion years of Earth's history has passed by, and all that biology has to show for it are single cells occasionally getting into groups to form colonies and stromatolites.

Then suddenly (in the sense of geologic time) life goes crazy. In a short burst lasting around 25 million years, starting about 541 million years ago, most of the major phyla of animals appeared. This great blooming of new species is known as the "Cambrian explosion." Humans could not breath or survive on Earth until around 500 million years ago – we would not exist if other life forms had not pioneered the way to make Earth habitable for us.

One thing we must get straight is that the land was colonized by life that first evolved in the ocean. The first simple fish, a jawless eel, appeared in the ocean around 510 million years ago, long before there were any true land-based animals. The first plants on land did not arise until around 450 million years ago. The first fish with legs, *Tiktaalik*, did not appear until around 380 million years ago. It is not until the Carboniferous, from around 360 million years ago, that Earth begins to look familiar, with corals, sponges, fish, and sharks in the ocean and insects, flowering plants, and tree ferns on land.

The Carboniferous was an amazing period in Earth's history. Giant tree ferns and tropical rain forests flourished after the evolution of plants on land, and this caused a spike in oxygen, rising to as high as 35% of the atmosphere. This extra oxygen allowed Carboniferous insects and reptiles to attain gigantic sizes – dragonflies with 75 cm (two and a half feet) wingspans and 30 cm (1 foot) wide spiders. Coal deposited during the Carboniferous fueled the Industrial Revolution in Europe. Climate change related to the formation of Pangea and the formation of polar ice caps ended the Carboniferous and oxygen levels fell back to around 20%, and have stayed at that level up to the present time.

Why does the atmosphere currently contain around 21% oxygen and not 10% or 30%? The Gaia hypothesis of James Lovelock and Lynn Margulis proposes that Earth's oxygen levels are maintained "by and for" life. In other words, the level of oxygen is 21% because that is the value which is balanced by self-regulating life processes on Earth (the state of homeostasis). In fact, scientists are not certain exactly why the atmosphere has maintained this percentage for the last 300 million years. It is a mystery.

What took life so long to get started? Why the sudden "explosion" of new species in the Cambrian? Well, maybe there wasn't an explosion at all. Maybe the fossil record just makes it look like one.

The Cambrian explosion filled the oceans with a diversity of life: trilobites, echinoderms, crustaceans, cnidarians, and others, the ancestors to all the major groups of animals. But evidence is growing that there may have been Precambrian precursors to many of these phyla. Maybe what appears like an explosion in species numbers is actually an artefact of what fossils were preserved. Because the simple truth is that everything we know about life in the geologic past is derived from fossils preserved in sedimentary rocks. And the fossil record is anything but complete. It has many gaps and holes with big blocks of time missing altogether. We need to be careful how we interpret what we find in the fossil record.

***

According to the Census of Marine Life, there are at present about 250,000 known species in the ocean.[6] The actual number is a mystery, but it could be as high as 1 million. The species alive today are a small subset of all those that have ever existed, perhaps only 1% (99% of species that ever lived are extinct). On average a species typically survives between 1 and 10 million years. Fossils are the only testimony of the existence of species that have gone extinct over geologic time, and as noted by Charles Darwin, the geologic (fossil) record is far from perfect.[7] Paleontology is the branch of geology devoted to the study of fossil life; the paleontologist slogan is: "when they're dead, they're ours!".

What is the probability that the existence of a species will be recorded in the fossil record? The answer is: it depends.

Consider a jellyfish.

It has no hard body parts and when it dies its body rapidly disintegrates. The chances of finding a fossil jellyfish are therefore very poor indeed. By contrast, a snail has a hard shell that is easily preserved, and geologists have found many snail fossils; this simple fact explains to some extent how the Natural History Museum in London has come to own 6 million fossils of different marine snails. Consequently, we have better records of species that have hard body parts (corals, snails, and the bones of fish or mammals) than species that have no or few hard body parts (jellyfish or octopus).

Chances of preservation are improved for species that have a widespread distribution across the Earth. Animals that have a very limited spatial range (live only in a restricted geographic area) are less likely to make it into the fossil record. Bigger organisms are also more likely to be preserved than smaller ones.

Another factor is the sedimentary environment where an animal dies and the likelihood of its body being buried before it has decomposed. The chances of preserving a dead body that gets incorporated into a river delta are much higher than one washed onto a rocky shore. If there is not a rapid deposition of sediment to bury the body before it rots away, there will be no fossil. The thickness of a bed (sediment layer) is no guide to the length of time required to deposit it. A 10 cm (4 inch) bed could represent a layer deposited in a matter of minutes in the case of a slump or debris flow. Tidal currents can lay down a bed within one 12-hour tidal cycle on the slip face of a migrating tidal sand dune. A storm event or volcanic eruption can also lay down a thick bed of sediment within a few hours. By contrast, a centimeter of mud could take 10,000 years to be deposited in the deep sea where only the slow rain of sediment falls from above.

Bed thickness per unit time ranges over 11 orders of magnitude. Fossil preservation is increased in rapidly accumulating sediment deposits – the faster the better!

Finally, there is the (small) probability that the deposit will itself survive ongoing geological processes and remain intact until the present time. Sedimentary deposits are themselves only temporary resting places for sediments (and fossils) on a longer journey. For example, sediments deposited in a river point bar may only survive a few years or decades until the river meanders again and the deposit is eroded and the

---

[6] http://www.coml.org

[7] Kidwell and Holland (2002).

sediment is transported further downstream. In an estuary or river delta, deposits may remain undisturbed for many thousands of years. But then an episode of lower sea level will expose them to the erosive forces of rain and flowing surface water. They will be eroded and redeposited in a new estuary or delta located further seaward or else exported to the deep sea, along with any fossils that they may have contained. What are the chances that the fossil will survive this second cycle of exposure on the surface and reburial? Or a third or even a fourth cycle?

Sediments are deposited and eroded and deposited over and over in natural systems that we can observe today. They are eroded by storms, underwater landslides, and glaciers, transported a few meters or hundreds of kilometers before they are deposited again. The chances that a shell or bone will survive these deposition-erosion cycles intact are very small indeed.

Deposits that might last a few million years are eventually carried by continental and oceanic plates across the globe. Most sediments deposited in the ocean are eventually subducted into the mantle. Only a few oceanic deposits are "lucky" enough to get scraped off onto a continental plate, where they may survive for a time before they are in turn eroded and returned to the ocean by rivers, glaciers, and wind.

The deposits and fossils that we see today exposed in layers of sedimentary rock are like the records kept by an absent-minded bookkeeper, who has left pages of his notebook strewn over a desk. Each page gives a glimpse into the past for a brief span of time. Working out the order that the pages should be in is sometimes difficult to figure (our bookkeeper didn't bother to number them!). There are many pages missing, more from ancient times than recent. The history of the Earth is written in a book where we have only one in ten pages intact, and the remainder are in an uncertain order.

The fossil data that do exist show that the Cambrian explosion of life was more like a blooming from an already fertile pool of phyla. Evolutionary radiation in the early Cambrian coincides with many environmental changes. It came immediately after the Earth's second major snowball phase (Cryogenian, 720–635 million years ago). There followed an increased oxygenation of the oceans, an atmospheric ozone layer was established filtering UV radiation, and there were changes in ocean chemistry, particularly increased calcium carbonate concentrations (needed by many marine organisms for building shells and hard body parts). Recent research indicates the explosion of species was linked to an 80% increase in atmospheric oxygen.[8] Changes in these environmental variables opened new niches and evolutionary pathways for organisms.

The records left by the absent-minded bookkeeper for the last 600 million years are in much better shape than the earlier parts of the story. Over this time, the oceans have been filled with living things that have left a fossil record, and our knowledge of the movement of the continents is far better. We can now tell the story of how the ocean basins evolved.

***

---

[8] Edwards et al. (2017).

Let's recap the story of our oceans so far. The water that fills the oceans today was probably formed around 4.2 billion years ago. Water completely covered the planet back then, apart from volcanic islands and possibly some small continents. Plate tectonics got started, and the creation of continents began. It took 2 billion years before we had the first 40% of the continents and the remaining 60% of continental area was created in the last 2 billion years. Through all of this time, ocean crust has been created along the mid-ocean spreading ridges and destroyed as it is subducted back into the mantle. This means that, whereas the continents are billions of years old, the ocean floor is merely 80 million years in age on average. How strange that the ocean basins are so very young. Much younger than the 4.2 billion-year-old ocean waters they contain.

Ocean basins are defined by the arrangement of continents at any particular time, as they are rifted apart and then smashed together again by plate tectonic forces. There are presently seven major plates which together cover 94% of Earth's surface, plus numerous smaller plates (and microplates – around 52 in total). Each of the different oceans we have today is at different stages in their evolution according to the "Wilson cycle" of the birth and death of oceans. The evolution of an ocean basin begins with the rifting apart of a continent – different "oceans" are recognized based on the continents that define their boundaries.

The great rift valley of Africa (where human's evolved) is an example of a continental rift that will eventually evolve into a new ocean (Fig. 3.3). As rifting pro-

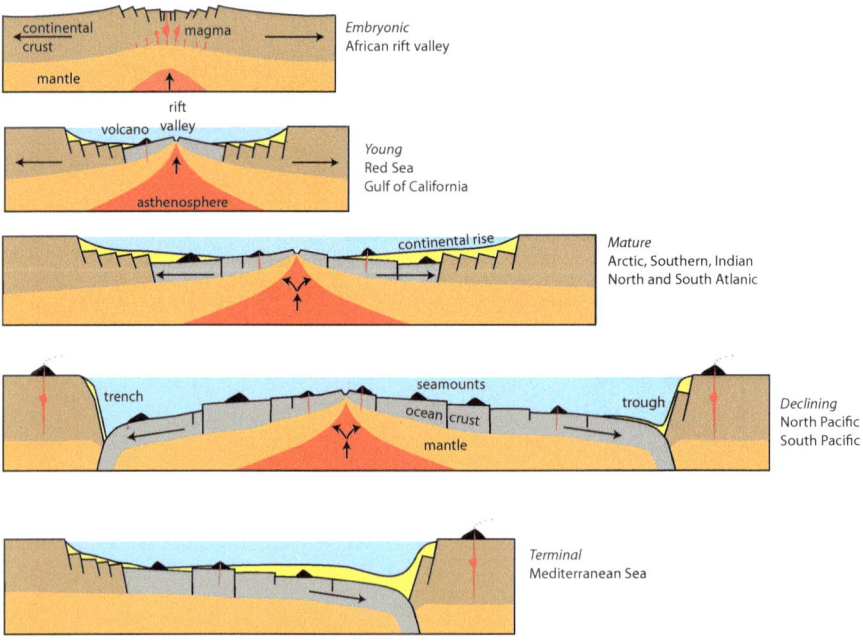

Fig. 3.3 Evolution of ocean basins according to the "Wilson cycle". (After Harris and Macmillan-Lawler 2017)

gresses, the opposite walls of the rift valley move further and further apart, while the valley floor subsides. Over time the valley floor will fall below sea level, and the rift valley will become flooded with seawater.

The Red Sea and the Gulf of California are examples of rift valleys that have reached this "young" evolutionary stage. In these places the high-riding, rifted continental plates are separated by lower elevation valley floored with ocean crust that seawater has flooded. The average age of ocean crust in these two locations is less than around 8 million years. The opposing rift valley walls will eventually become parts of two separate continents, referred to as "conjugate" continental margins. The east coast of South America and west coast of Africa are good examples of conjugate continental margins that inspired Alfred Wegner's continental drift theory.

The Arctic Ocean contains a young spreading ridge that has split Norway away from Greenland as well as knocking off a chunk of northern Russia, forming the Lomonosov Ridge that has been left stranded below the North Pole. The so-called Gakkel spreading ridge that lies between the Lomonosov Ridge and the Russian mainland is the northernmost extension of the mid-Atlantic ridge, passing northward from Iceland and zigzagging northward of Svalbard into the Arctic Ocean. This part of the Arctic Ocean basin is less than 80 million years old.

As the mid-ocean ridges push the continents apart, some pieces are forced into awkward spaces that cause them to split and crack apart. The boundary between the North American plate and the Asian plate is a good example: where do you suppose the North American plate ends and Asian plate (Russia) begins? If you guessed that the boundary must be in the Bering Sea, between Alaska and Siberia, then sorry, but you're wrong!

The plate boundary between North America and Asia is actually much further to the west in Siberia. It extends through Siberia and into the North Pacific before curving around to join the deep ocean trench adjacent to the Aleutian Islands (the Aleutian Trench; Fig. 3.4). From a plate tectonic perspective, this is one of the least well-understood plate boundaries on Earth.[9] Nevertheless, the North American plate includes all of the shallow Bering Sea along with the eastern part of Siberia. It also includes an ancient remnant ocean basin adjacent to the north slope of Alaska and Canada, known as the Canada Basin, that is at least 160 million years old. Hence the Arctic Ocean is a conglomerate of young, newly formed ocean crust adjacent to Svalbard in the east and the ancient Canada Basin in the west.

As conjugate plate margins move further apart and the ocean basin widens, sediments from the adjacent land spill into the basin and sediment layers typically several kilometers thick are deposited. The sediments come from the erosion of mountains by rivers and glaciers, and they all eventually reach the ocean, transported by water, ice, and wind. The Arctic, Southern, Indian, South Atlantic, and North Atlantic Ocean basins have reached a "mature" evolutionary stage, and they all contain such thick sediment layers (Bruce Heezen termed these as "Atlantic"-type margins). The *average* sediment thickness in the North Atlantic Ocean is over

---

[9] Fujita et al. (2004).

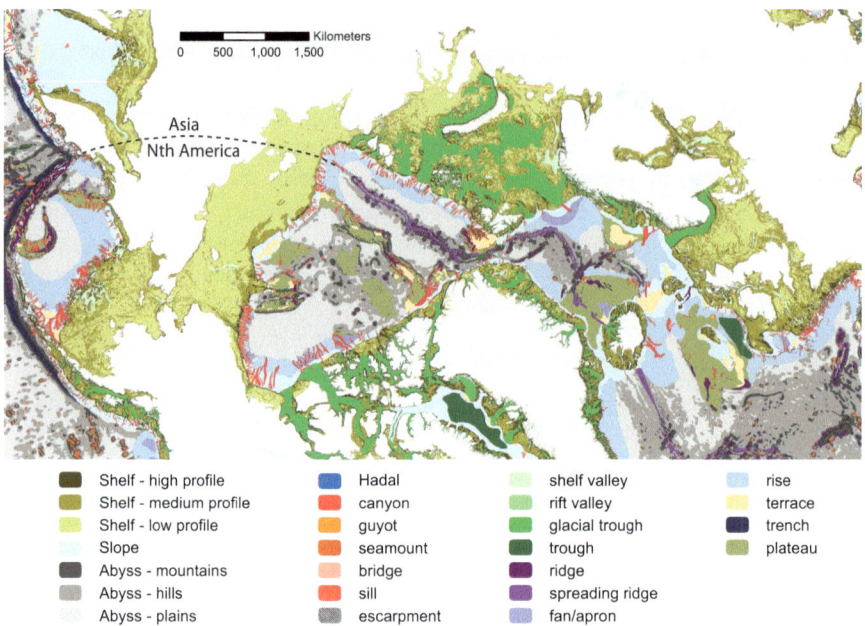

Fig. 3.4 Geomorphology of the Arctic Ocean viewed in North Polar Projection, modified from Harris et al. (2014) to illustrate the location of the plate boundary between Asia and North America

1400 m. Where it is piled up against the continental margins, sediments in some places are over 4 km thick. Sediment is much thinner (or absent) over the mid-ocean ridge.

Throughout this evolutionary process, ocean crust is formed along the mid-ocean spreading ridges (where the ocean crust has an age of zero), and the oldest ocean crust is located adjacent to the sides of the two conjugate plate margins (buried beneath the thickest sediments). The age of the oldest ocean crust (adjacent to the rifted conjugate margins) is an approximate indicator of when these mature ocean basins started to form. And what we discover is that the Southern, Indian, South, and North Atlantic Ocean basins formed between 140 and 180 million years ago. This was when Gondwana and Laurasia began to disintegrate into the continents we have today.

The next phase in the ocean evolutionary cycle is the commencement of the subduction of ocean crust, which requires the formation of ocean trenches. Oceans acquire deep trenches as they mature. Their development is a gradual process. The percentage area of trench is actually a good indicator of the transition of an ocean basin from the "mature" to the "declining" evolutionary phase. Like the wrinkles on your face, the more trench an ocean has, the greater its maturity and closer it is to a declining stage of life.

At the present time, the Arctic Ocean has no ocean trenches (the Arctic Ocean is just a kid really). Similarly, the Southern Ocean is only 0.04% trench which is the

South Sandwich Trench located north of the Antarctic Peninsula. The Indian Ocean has 0.2% trench, which includes the Java Trench located adjacent to Indonesia. The South Atlantic Ocean has 0.2% trench, and 0.3% of the North Atlantic Ocean is ocean trench (the Puerto Rico Trench).

In all mature oceans, the rate of ocean crust production is much greater than the rate at which ocean crust is destroyed in subduction zones. Mature oceans are expanding, and their trenches are just getting started!

The opposite is the case for ocean basins that have entered the "declining" phase; the South and North Pacific Oceans contain about 1% ocean trench by area. These include the oceans major trench systems: the Chile Trench, Aleutian Trench, Japan Trench, Kermadec Trench, and the Mariana Trench. In the Pacific, the rate of ocean crust destruction is greater than the rate at which it is being produced. The North and South Pacific Oceans are slowly shrinking in size.

In the declining phase, oceanic crust is still the oldest next to the continents where it is being subducted into trenches and down into the Earth's mantle. But the age of the ocean crust is not related to the age of these "old timer" ocean basins, because the oldest crust that was created when these oceans were first formed has long since been destroyed by subduction.

The subduction zones also capture much of the sediment that comes off the continents and carries it directly down into the mantle for recycling. The result is that sediment thicknesses in declining ocean basins, like the Pacific Ocean, are on average only a few 100 m. Subduction along the tectonically active Pacific margins also gives rise to massive earthquakes and the eruption of volcanos around the so-called Pacific Ring of Fire.

The rate of seafloor spreading is faster in "declining" ocean basins than it is in "mature" basins. The reason is probably because in mature ocean basins, the ocean crust that is formed at the mid-ocean ridge is pushing the conjugate continental plate boundaries apart so the spreading rates are slower, typically around 20–30 mm/year. These are the so-called passive margins of the Arctic, Atlantic, and Southern Oceans. In contrast, seafloor spreading rates in declining ocean basins have the combined benefits of the mid-ocean ridge "push," where ocean crust is being formed, plus the "pull" from the sinking of ocean crust back into the mantle. Seafloor spreading rates in the Pacific are fast and average around 57 mm/year.

As an ocean basin shrinks, it is squeezed out of existence by converging continents that are barreling toward each other like an unstoppable train wreck. The Mediterranean Sea is the last remaining piece of the once mighty Tethys Ocean, and it is being squashed out of existence by the collision of Africa and Europe. The collision has already created the Alps and the Pyrenees mountains which we might expect 1 day to grow as large as the Himalayas. Sediments, eroded from these mountains, have filled in the Mediterranean basin to a significant extent, averaging over 4 km in thickness. The last phase of an ocean is thus marked by colliding continents, the formation of new mountain chains, and filling in the remaining ocean basin with sediments.

And now we have come to the end of the beginning of our story. The evolution of the oceans has a 4-billion-year history following the collision of planet Theia with the Earth. Volcanic eruptions and meteor impacts produced the early atmosphere from which the oceans condensed as the Earth cooled. Plate tectonics produced the continents, which are the by-product of the melting and distillation of ocean crust as it is subducted back into the Earth's mantle. Continents divide the ocean into basins of varying size, which constantly change as the continents wander over the globe, occasionally combining into supercontinents before dispersing again.

Continents are moved by the seafloor-spreading conveyor belts, as ocean crust is created at mid-ocean ridges before eventually being destroyed in subduction zones. Ocean crust we see today is very young, around 80 million years on average, whereas the water within the basins is nearly as ancient as the Earth itself.

The oceans and continents that we have today are at various stages of the Wilson cycle. This understanding has taken over 100 years of scientific exploration and discovery to achieve. It has taken all the marine geologists who lived prior to the 1960s to put together the theory of plate tectonics, and the generations that have followed have refined and tested the theory. But we are still a long way away from having a complete picture of how all the pieces fit together. Partly this is because we still have not completed the collection of some very basic data sets. Take, for example, the depth of the ocean.

Chapter 4
How Deep Is the Ocean?

"This new (sonar) method has in a few years completely altered our concept of the topography of the ocean bottom. Basins and ridges, troughs and peaks have been discovered, and in many areas a bottom topography has been found as rugged as the topography of any mountain landscape. "
Harald Sverdrup, Martin Johnson, and Richard Fleming
The Oceans, 1942

Abstract How deep is the ocean? To find the answer to this question is not as simple as you may think. Ocean mapping technology has developed over time, and there have been many advances like the invention of sonar. But it is a fact that Mars is better mapped than of our own ocean floor. The search for the missing Malaysian Airline Flight MH370 in the Indian Ocean illustrates how poorly our oceans are mapped. In this chapter we will learn how to take a piston core sample of the ocean floor, and we will meet Maurice (Doc) Ewing and learn how he invented seismic profiling so that we could "see" the layers of rocks and sediment beneath the ocean floor. But all of our sonars and seismic systems make a lot of noise. How has this impacted on marine mammals and other ocean creatures?

Keywords Maurice (Doc) Ewing · HMS *Challenger* · Mariana Trench · Sonar · Dead reckoning · Multibeam echosounder · Malaysian Airline Flight MH370 · Seafloor mapping · Börje Kullenberg · Piston core · Seismic profile · Hydrophone

Perhaps the biggest mystery of the ocean is how to study it in the first place. The main reason is that the ocean is opaque – we cannot see into it more than a few 10's of meters even in the clearest water and much less than that in most places. Light does not penetrate easily through water. Below a depth of around 200 m, there is no light at all. Most of the ocean floor and hence most of the Earth's (submerged) land surface is in total, perpetual, darkness. It has never seen sunlight, and it never will.

© Springer Nature Switzerland AG 2020   41
P. T. Harris, *Mysterious Ocean*, https://doi.org/10.1007/978-3-030-15632-9_4

The oceans appear deep to humans, but "depth" is a relative term. In proportion to the 12,742 km (7917 miles) diameter of Earth, the 4 km deep ocean is a thin skin on the surface. If Earth were a sphere 3 m in diameter, the ocean would be on average only 1 mm thick. On the other hand, the depth of the ocean is large at the scale of Earth's surface features. If Mount Everest – the tallest mountain on Earth – were transported to the bottom of the Mariana Trench, there would still be 2076 m (6811 feet) of water above its peak. If there were no plate tectonics and the Earth's surface was flat, the oceans would cover the world to a depth of about 2600 m. No land would mar the pale blue surface of planet "ocean," and all life that evolved would be aquatic.

But we do have continents, and they divide the single world ocean into smaller pieces of ocean that we have given names: Atlantic, Pacific, Indian, Southern, and Arctic. For most of human history, maps of the oceans were simply their outlines as determined by the shapes of the continents. On early maps, the oceans were depicted as blue-shaded, blank, nothingness. Worse, they were a place of sea monsters and other unknown dangers, a seemingly unknowable part of our world that would remain forever a mystery. Humans were ignorant of what lay beneath the ocean surface until only the last century when ocean exploration finally began.

H.M.S. Challenger embarked from Portsmouth, England, on December 21, 1872, on its 3.5-year expedition to begin the scientific exploration of the oceans. When it finally returned home on May 24, 1876, the Challenger had zigzagged around the globe a distance of 68,890 nautical miles and had visited every continent, including Antarctica. In order to measure the ocean depths, the Challenger was equipped with 144 miles of sounding rope together with 12.5 miles of piano wire for the deployment of bottom sampling gear.

During its voyage, *Challenger* would "heave to" at uniform time intervals, and a series of standard observations were made to obtain a measurement of depth and to retrieve a sample from the seafloor. A total of 362 observing stations were completed on the voyage, and 492 separate depth soundings were made, the first time that anyone had sampled the deep seafloor in a systematic manner. The Challenger expedition revealed the first broad outline of the shape of the ocean basins, including a rise in the middle of the Atlantic Ocean that we now know as the Mid-Atlantic Ridge. The deepest parts of the oceans were not located in the middle, as you might think, but along some of the edges, near to land.

The deepest place in all the oceans is near where the Challenger took one of its 492 soundings. This location is a section of the Mariana Trench, named the Challenger Deep, and it is 37,800 feet (about 11,000 m) deep. What are the chances that Challenger would sample the deepest ocean trench on Earth with one of its 492 randomly placed samples? Just how lucky the crew of Challenger was can now be realized. We now know that out of 362 million km^2 of ocean, there is only about 2 million km^2 of deep ocean trench. The Challenger therefore had a 0.55% chance of sampling a trench at each of its randomly placed stations. But the chances of Challenger randomly choosing the deepest, 1 square kilometre in the ocean for one of its stations are one in 362 million!

The task of mapping the ocean floor must have seemed impossible to these early explorers. Each of the 492 deep ocean soundings would have taken the best part of a day to complete. The deeper the water, the longer it took to pay out the wire, meter by meter, watching the big winch wheel turning slowly as the wire went down. Hours later, a weight on the end of the wire would hit finally the ocean bottom, taking off some of the wire tension, measured using a tension gauge (dynamometer). It is no simple thing to know when the weight has hit the seabed. If your wire weighs 1 kg/m length, for example, then each kilometer of wire weighs 1 ton. In water depths of 2 or 3000 km, the weight of wire paid out would be several tons. A 100 kg lead weight at the end of 3 tons of wire does not release that much weight in proportion to the total tension when it hits the bottom, so you need a fairly sensitive tension gauge attached to the winch to make sure you know when the weight does hit bottom.

The depth is estimated by how much wire was paid out, not an exact measure if there is any angle on the wire caused by currents or drifting of the ship while it's been on station. Therefore, the depth is corrected by taking a second measure of wire angle. It must be assumed that the wire is in a straight line, not curved, a further source of error.

Then comes the slow, painful process of reeling the wire back up again, hour after hour. Challenger had a steam-powered winch on board; otherwise any deep ocean soundings would have been nearly impossible. When at last the weight came into sight and was pulled back aboard, the ship could move on to the next station. The bottom of the lead weight was covered in wax to trap a small sample of whatever mud or sand lay on the seabed.

Dropping a weight at the end of a rope was the only way to measure depth for most of maritime history ("by the mark twain" was the call of leadsmen to the captains of Mississippi riverboats to announce two fathoms of water depth). Lucky for oceanographers, taking accurate depth measurements is possible nowadays by other, much faster methods. Thanks to the physics of water, sound travels very effectively through the depths at a relatively constant speed of around 1500 m/s (the speed varies with water temperature and salinity). Water depth can then be measured by making a loud noise and counting the seconds before you hear the echo. Then simply multiply the elapsed time by the speed of sound, divide by two (to allow for the travel time from the ship to the seafloor and back again), and presto! You have a quick and accurate estimate of the water depth!

It was not until World War I that the first working electronic echo sounders were invented. The technology, with the acronym "ASDIC" (named after the Anti-Submarine Detection Investigation Committee), was kept a secret at first because it was used to detect submarines. Later the technology became known as SONAR (originally an acronym for SOund Navigation And Ranging). An echo sounder or sonar device emits a sound pulse which bounces off the seabed (or a submarine, if present) and records the depth at a single point below the ship's hull.

Like headlights on a car, echo sounders are now standard equipment on all ships and boats for safe navigation (knowing there is plenty of water below the keel makes your captain happy!). Using modern echo sounders over the last 50 years, sailors and scientists have now recorded more than 290 million ocean soundings over the oceans, and our maps of the seafloor are getting pretty good.

Pretty good, but not that good.

Now, you would think that 290 million depth soundings would be enough to make a fairly good bathymetric map and keep oceanographers happy, wouldn't you? But when you recall that the ocean covers an area of 362 million km^2, that makes less than one sounding per square kilometer on average. And when you realize that most of those 290 million soundings were taken close to the main seaports and along the edges of the continents to find things that ships are prone to bump into by accident, it turns out that there are areas of ocean the size of Texas that do not have a single depth sounding.

Then there is navigational error. Each depth measurement is only as useful as the extent to which you know where the ship was on the ocean when it was taken. Prior to the widespread use of the global satellite positioning system (GPS), available since the early 1990s, most ships had only dead reckoning supplemented with occasional satellite fixes, and their navigational accuracy was probably plus or minus one nautical mile (on a good day). Many of the Challenger's soundings are probably located with a positional accuracy of several miles, when you take into account that navigation was based on dead reckoning[1] with a noon sighting of the sun (for latitude) and chronometer reading (for longitude) as the main navigational methods. So, if we exclude soundings that could be biased by inaccurate navigation, in reality we have had only the last 20 years or so to collect accurate ocean sounding data.

The good news is that our modern echo sounders are much improved. Single beam echo sounders that measure just one spot directly below the ship's hull have evolved into multibeam echo sounders (Fig. 4.1). These new devices can simultaneously measure 100 or more spots across the track of the ship, directly below and off to either side of the ship's hull. The width of area mapped is equal to about five times the water depth for most multibeam systems. This means that in water that is 2000 m deep, a typical multibeam echo sounder can measure depths in a 10-km-wide swath, with soundings spaced 100 m apart, and at a positional accuracy of plus or minus a few meters.

If our goal is to map the whole ocean using multibeam echo sounders, even with 1 km resolution, we have a lot of work to do. Ship-based sonar and other measurements have mapped such a small percentage of the ocean floor that it would take a single ship 120 years (or 10 ships 12 years) to measure all the ocean floor depths, according to published US Navy estimates. In an exciting program, the Nippon Foundation and the Guiding Committee of the General Bathymetric Chart of the Oceans (GEBCO) announced their intention of mapping 100% of the deep ocean

---

[1] "Dead reckoning" is a term sailors use to describe the process of using the ship's speed and course to estimate its present position by extrapolating backward to the ship's last known position. This works fine if there is no current or wind shear blowing you off course (except there is nearly always some current and wind shear!).

Fig. 4.1 Multibeam echo sounders can map the seafloor in a swath across the track of the ship to produce an accurate image of the seabed. (Image courtesy of New Zealand Institute of Water and Atmospheric Science (NIWA), copyright 2005. All rights reserved. https://www.niwa.co.nz/services/vessels/niwa-vessels/rv-. This is called "swath mapping" because the ship travels back and forth along parallel lines, like mowing the grass, mapping swaths of seafloor)

floor at a resolution of 100 m (300 feet) by the year 2030.[2] It is estimated that the cost of completing this mapping effort would be around US $3 billion, about the same cost as one of NASA's Mars space probe missions.

Since multibeam echo sounders can only map a swath that is about five times as wide as the water is deep, it actually takes longer to map in shallow depths. This is because if it is only 100 m deep, for example, the swath width is reduced to 500 m, and if it is 10 m deep, the swath is reduced to 50 m, etc. So, once we have mapped the deep oceans, it would take another 750 ship-years to map the world's continental shelf.[3]

Before we started using multibeam sonar, there was a general perception that most of the ocean floor is a barren wasteland, a featureless, flat landscape of abyssal plains extending to the submarine horizon, interspersed with monotonous, mud-draped hills and valleys. But in places where we have mapped the ocean floor with multibeam sonar, a completely different image is revealed. What we have found is that the ocean floor is as complex and varied as the terrain we see on land. Sometimes the seafloor features are subtle, and their character is only truly revealed when

---

[2] https://www.nippon-foundation.or.jp/en/news/articles/2016/33.html

[3] Becker et al. (2009).

viewed at a high enough resolution; in other places the features are as dramatic and rugged as any mountains or canyons seen on land (more on this later).

Application of the new sonar technologies has provided a patchwork of well-mapped areas of seafloor, mainly along shipping routes and around areas of interest to the offshore petroleum industry, draped over poorly mapped seafloor. Most of the ocean floor is so poorly mapped that you could easily lose a large passenger aircraft among the vast areas of unmapped, unexplored seafloor terrain. And in 2014, that is exactly what happened.

***

On March 8, 2014, while flying from Kuala Lumpur to Beijing, Malaysian Airline Flight MH370 was lost at an unknown location somewhere over the southern Indian Ocean. The search that followed, which primarily involved the Malaysian, Chinese, and Australian governments, became the largest and most expensive in aviation history. After some initial confusion over the region where the plane was supposed to have been lost, the search was directed to an area of the southern Indian Ocean located about 1800 km (1100 miles) south-west of Perth, Western Australia. This area is among one of the least explored parts of the ocean.

Since the exact location of the MH370 crash site was (and still is) unknown, the search area was only broadly defined as an elongate swath about 200 km wide and 1500 km long, its curved shape representing the assumed flight path of the doomed aircraft. A daunting task awaited the searchers – they were looking for signs of airplane wreckage on the remote ocean floor, in places over 4000 m deep.

The searchers were hunting for the largest intact piece of aircraft that could possibly survive impact at great speed with the surface of the ocean. The most likely remains are the Rolls-Royce Trent 800 series turbofan engines used on the Boeing 777s, which are 2.4 m (8 feet) in diameter and 4.5 m (15 feet) in length. To resolve an object of this size would require sonar resolution of at least 0.5 m (18 inches), if not better. Larger pieces of the plane's fuselage might also have reached the ocean floor, but it seems doubtful that the searchers would use a resolution of less than 1 m to be confident of being able to detect any wreckage.

In order to achieve the necessary high (<0.5 m) resolution, the searchers used towed side-scan sonars and autonomous submarine vehicles with built-in high-resolution sonar equipment and magnetometers to detect metallic aircraft components. These devices needed to reach closer than 100 m to the seabed in order to be able to detect objects of this size. But the existing maps of the seabed are so crude that the depth was not known better than ±100 m. This could result in valuable equipment being towed into the side of an unmapped hill or ridge.

The solution was to first map the search area using ship-mounted sonars to make a new base map, with good enough resolution to deploy the expensive deep-towed and autonomous (robot) equipment. Three ships worked for 2 years, between June 2014 and June 2016, to collect the base map data over an area of 279,000 km², an area larger than the US state of Oregon. The deep-towed survey equipment was deployed as soon as an area had been mapped by surface vessels, allowing the search to continue in parallel. In late 2017 the Malaysian government signed a deal

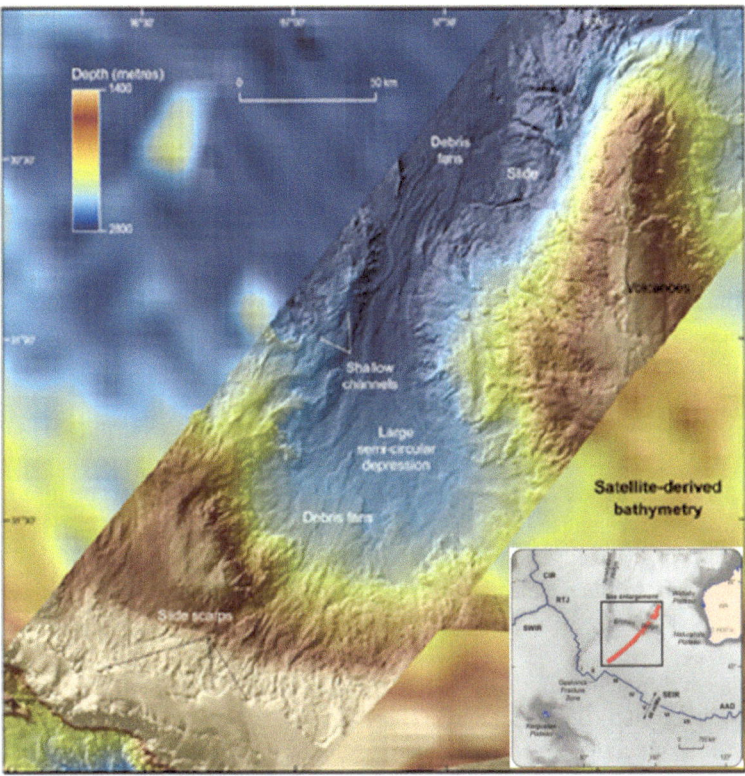

Fig. 4.2 Portion of the area mapped using multibeam sonar in the search area for lost aircraft MH370 in the southern Indian Ocean. The image contrasts the low quality of the existing satellite-derived bathymetry with the higher-resolution image produced using multibeam sonar

with US-based survey company, *Ocean Infinity*, to continue the search using six autonomous survey vehicles in the mapped area.

Although the search for the wreckage of MH380 did not find any sign of the aircraft on the ocean floor, the data collected have revealed how poorly mapped our ocean floor actually is. This can be clearly demonstrated by comparing the old and new seafloor maps (Fig. 4.2). The new map agreed with only 62% of the old map within an error margin of ±100 m. The old map was wrong by over 100 m depth in the other 38% of area mapped. The worst error was a 1900 m (6230 feet) tall seamount that was discovered on the new map that was not shown on the old map. In fact, the new data revealed 15 previously unknown seamounts along with escarpments, ridges, and canyons that were also unknown to exist.[4]

In short, seafloor mapping carried out in the search for MH380 has demonstrated the poor quality of our existing map of the ocean floor. It is sadly true that we have better maps of the surface of Mars (4 m pixel resolution) than we have for our

[4] Picard et al. (2018).

oceans (far worse than 1 km resolution in most places). If you have ever been on an aircraft and flown over the ocean, you will have seen the digital seafloor map[5] that I am talking about. It's the one used to show the in-flight track of the plane. This is the map we must use to search for any aircraft or ships lost in the ocean. It is the map used by oceanographers to model tsunamis and to make predictions for how ocean currents will change due to global climate change. It's the map used by engineers to select routes for laying submarine cables, by geologists to study earthquakes, and by submarine commanders to avoid running into the seabed (on January 8, 2005, the US nuclear submarine *San Francisco* ran into an uncharted seamount located approximately 360 miles southeast of Guam and narrowly avoided sinking). Errors and inaccuracies in our existing bathymetry map reduce the reliability of scientific predictions and impact on all human activities that depend on knowing the depth and shape of the seafloor.

Creating a new, better map of the ocean floor will be a task for the next generation of marine scientists. There are literally 1000 undiscovered mountains, ridges, and canyons, inhabited by unknown species of animals, waiting to be discovered, mapped, described, and sampled.

If measuring the depth of the ocean is hard to do, imagine how hard it is to collect a sample from the deep ocean floor?

***

Deep environments are hard to get to, being located on the seafloor in water that is several thousands of meters deep. There are several ways to collect seabed sediment samples: submersibles and tethered remotely operated robots are examples. But the simplest (and least expensive) ways generally involve lowering a sampling device on a wire over the side of a ship to try and grab a sample of seafloor mud and hopefully a few specimens of the creatures living in it or on it. Though it sounds easy, this is actually a rather tricky thing to accomplish.

The most common device used to obtain a core sample of seabed sediment is called a piston corer. A core sample is preferred over a surface sediment grab for obtaining a vertical profile of sediment layers to get a complete "book," with all the pages intact and in the right order. It is easier to collect a grab sample using a device that usually collects a mixed-up mush of sediment lying on the surface of the seabed (Fig. 4.3a). This is often all that is needed to collect a representative sample of surface sediment and is also used to collect biological specimens. But a grab sample is not useful for collecting an undisturbed profile of the sediment, which is necessary to keep the sediment layers and the story they have to tell, in chronological order.

Getting a good core sample has preoccupied marine geologists over the last century. Many inventions have been tried and failed. One example is a coring machine invented in the 1930s by Dr. Charles Piggot from the Carnegie Institution, which included an explosive charge that literally shot the core barrel into the seabed like a bullet! Another idea, also from the 1930s, was a vacuum Varney-Redwine core tube; on hitting the seabed, the tube ruptured and the core would supposedly be sucked

---

[5] Smith and Sandwell (1997).

Fig. 4.3 Examples of devices used for collection of seabed samples: (**a**) Shipek grab sampler, with stainless steel pan underneath in the closed position; (**b**) a hydraulic rock drill, used for collecting drill-core samples of soft sedimentary rocks like limestone; (**c**) a box core used to collect a volume of soft sediment to a depth of around 50–70 cm; (**d**) a piston corer assembled on deck with trigger arm attached and ready for deployment. (Photos taken by the author)

into the seabed by the vacuum pressure. Other devices used include rock drills (Fig. 4.3b) and box corers (Fig. 4.3c).

The piston corer, by comparison, is a simple mechanical device that includes a long pipe with a lead weight attached to the top (Fig. 4.3d). Inside the pipe there is a plastic core liner in which the core sample is saved. At the business end of the pipe, a sharp-edged "core cutter" is attached to ease the passage into the mud.

It is important to get a sample with the top-most sediment layer intact. This top layer is the stuff that was just recently deposited (and it's always nice to start any novel from the first page!). Preserving that top layer of sediment is where the "piston" comes in. It goes inside the core liner, so that as the corer sinks into the seabed, the piston rises at the same speed, protecting the important top layer.

The piston corer also overcomes another critical problem of trying to get a sea-floor sample from a ship. The problem is that, on the ocean, most of the time, ships are moving up and down as they ride over the waves. If the ship is moving up and down, so is the core machine that is being lowered at the end of the wire attached to the ship. Both ship and corer are moving up and down! If, while being lowered, the corer were to touch the seabed just as the ship is rising on a wave, it could cause the corer to be left standing impotently on the seabed in which case it will simply fall over and not take any sample at all! Or even worse, it could cause the corer to hit the bottom, rise up, and hit a second time. Then you would get two core samples from the same place stacked one on top of the other inside the core liner (the first few pages of the book repeated twice!).

To overcome this problem, a few meters of wire are coiled up, and a trigger arm is fixed above the piston corer. Then a weight is attached to the trigger arm on a separate long wire so that it hangs down below the end of the corer. The whole device is lowered to the bottom. The weight hits the bottom first releasing the piston corer which free-falls a few meters before hitting the seabed while the wire uncoils. This means that, for a few seconds, the piston corer is decoupled from the movement of the ship and is unaffected by waves. The free-falling speed also gives the corer more force so that it can penetrate deeper into the muddy layers.

If the lengths of wire have been measured correctly, the instant that the corer hits the bottom, the wire attached to the piston becomes taut, and the piston provides a small amount of suction. This is the final and most important role of the piston. The suction eases the passage of the corer into the seabed, greatly increasing its depth of penetration into the layers of mud.

This ingenious device was invented over 70 years ago by a Swedish marine scientist named Börje Kullenberg. The piston corer has been adapted for use around the world and now includes giant piston corers 30 m (90 feet) in length, using essentially the same Kullenberg design.

Once on board, the core liner is extracted from the pipe and cut longitudinally in half. One of the more exciting moments in the life of a marine geologist is that first glimpse of sediment when the two halves of core are split in two. What will the sediments reveal about the history of this place in the ocean?

Core samples are regularly collected by oceanographic research vessels operating around the world. But the most famous of all seabed sampling programs has been running for 50 years under different names and using different vessels: Deep Sea Drilling Project (1968–1983), the Ocean Drilling Program (1985–2004), Integrated Ocean Drilling Program (2004–2013), and the International Ocean Discovery Program (2013–present). The current program uses the drillship *JOIDES Resolution*, whose name is an acronym standing for "Joint Oceanographic Institutions for Deep Earth Sampling."

The seafloor drilling program has successfully sampled every ocean basin and drilled at over 2000 sites to reveal the geologic history of the oceans. It has involved hundreds of geoscientists from almost every country. The drillship *JOIDES Resolution* uses the most sophisticated technology available to image and sample the ocean floor, including a hydraulically powered "advance piston corer," which is based to a significant extent on the Kullenberg design.

***

Twenty-eight million barrels of oil per day are extracted from the world's continental shelf sediments (worth USD $2 billion per day, or $750 billion per year, at US $75 per barrel). This is accomplished by more than 6500 offshore installations scattered around the Gulf of Mexico, the North Sea, and the Arabian Sea among other places. Around 200,000 people are directly employed in this industry, exploring for oil deposits, drilling wells, and operating the production equipment. And it is all made possible because 80 years ago, a young marine geologist was curious about the transmission of sound waves through deep sea sediments.

This story is about the invention of continuous seismic profiling and of the person who did much to pioneer its development: Maurice (Doc) Ewing (who we met earlier as the Founder and first Director of Lamont-Doherty Earth Observatory, with his staff Bruce Heezen and Marie Tharp). As is so often the case in science, no single person can be given all the credit for an innovation. But Doc Ewing stands out as one of the leading figures in the development of this technology which underpins the offshore oil and gas industry.

The idea of using seismic reflection by setting off an explosive charge and then recording the echoes on a geophone (seismograph) was invented in the early 1900s. By moving the point of the detonation away from the geophone step-by-step, the echoes reflect back off different layers in the subsurface, creating an image of the structure of the Earth. This technology was first used on land to explore for oil deposits in Texas in the 1920s. Doc Ewing had the idea of using this same technology to map sediments on the ocean floor.

Born in Lockney Texas in 1906, Doc Ewing graduated from Rice University with a PhD in 1931. He had experience working with seismic reflection to explore for oil in Louisiana, his summer job while still a graduate student at Rice. In October 1935, Ewing was given a chance to try out his idea of using dynamite and seismographs to map submarine sediments on board the Research Vessel *Atlantis*, a ketch-rigged, steel-hulled sailboat, 43 m (140 feet) in length operated by Woods Hole Oceanographic Institute. How delightful it must have been to work as a marine scientist in those days, aboard a sailing vessel!

The first experiments involved using a seismograph lowered to the bottom from *Atlantis* and an explosive "bomb" charge dropped from a whaleboat. A line of seismic data was collected offshore from Cape Henry, Virginia, to the edge of the continental shelf. The data identified a 4 km (12,000 feet) thickness of sediment overlying basement, matching the arrangement of sediment layers on the adjacent land, which Ewing correctly interpreted as proof that the continental shelf is the submerged extension of the continent.

Ewing's next experiment was carried out 2 years later in 1937 when he again had ship time on *Atlantis*. His idea was to measure the sediment thickness in the deep ocean. In the era before plate tectonic theory, it was believed that the oceans had received sediments from land since the Earth had formed, billions of years ago. Logically, there should be many miles thickness of sediment resting above oceanic basement close to the continents.

To measure this thickness, Ewing proposed to attach a seismograph and a long string of explosive charges onto a very long wire cable and lower it to the seabed. Then the explosives would detonate according to preset, clockwork timers. Unfortunately, the experiment was a failure for a number of reasons: the clockwork timers did not detonate, and the cable was dragged accidently which disturbed the seismograph and other problems.

Ewing did not give up his idea and he tried again in 1940. This time he lowered the explosive charges and seismographs to the seabed using clever floatation devices that would sink to the bottom and then rise to the surface after a salt block had dissolved to release a weight. Again, the experiment did not work.[6]

---

[6] Schlee (1978).

World War II put a stop to experiments with scientific seismic profiling, but very soon after the war had ended, in 1947, Ewing was at sea again this time mapping the sediments over the mid-Atlantic ridge. A new approach to seismic profiling was used. Ewing realized that there was no need to deploy the seismograph on the seabed or to send the explosive charges all the way to the bottom. Instead the explosives would be set off at the sea surface, and a new device, called a hydrophone, would be lowered into the water to record the seismic signal. The water column would simply be treated as another layer, resting above the seabed. At each station the ship would heave to, an explosive charge would be set off behind the ship, and a hydrophone would record the signal.

And this approach worked.

Ewing discovered from his seismic surveys that the sediment thickness in the ocean was much less than was expected. There was virtually none over the mid-Atlantic ridge, and although there were some thick deposits adjacent to the continents, the volumes did not tally with the amount of sediment that the oceans should have received since the Earth was formed. The answer to this apparent paradox was eventually provided by the discovery of plate tectonics and by Harry Hess in his 1962 paper "History of ocean basins," but in 1947 these missing sediments were a disconcerting fact that could not be explained. But the most important outcome of this work was the invention of ship-based seismic reflection profiling, which Ewing proved was feasible and incredibly useful for imaging the submarine geology of the oceans.

The oil industry quickly realized the value of marine seismic profiling, and technological advances of seismic surveying quickly followed. Systems were worked out so that the ship did not need to heave to for every shot, thereby allowing "continuous seismic profiling" (CSP). In the 1950s a gun was developed that used compressed air as a sound source instead of dangerous explosions of TNT. Explosives can still be useful to penetrate thick sediment deposits and image their internal structures. I was aboard the RV *Discovery* in 1983 in the Thermaikos Gulf of the northern Aegean Sea, with Professors Mike Brooks and Mike Collins, when we shot a seismic line using explosive charges. I joined the other PhD students in watching the technicians working on the back deck and the spectacle of plumes of water ejected skyward with each detonation.

In recent decades, the advent of computers has revolutionized the recording of digital seismic data. A single hydrophone has given way to multichannel seismic systems with hundreds of hydrophones in a single streamer, over 8 km (4 miles) in length, towed behind a survey vessel. And single streamers that can collect data in two dimensions (2-D) have evolved into multiple (from 8 to 16) streamers towed in a parallel array that collect three-dimensional (3-D) seismic data. Once in production, some oil fields are resurveyed to assess progress in draining the oil reservoir over time (4-D seismic).

Most people don't realize that crude oil production on land plateaued at around 65 million barrels per day in the early 1990s. Nearly all the growth in oil production since then (apart from some growth in the United States recently from onshore shale oil) has been due to growth in the offshore oil industry which now accounts for around one-third of total production. Total oil production was about 97 million barrels per day in 2015.

Today there are over 140 specialized seismic vessels working in the offshore oil industry engaged in seismic exploration around the world. The United States alone awards around 20 permits per year for seismic exploration, and many other countries have programs to actively explore and assess their offshore oil resources.[7] The development of continuous seismic reflection survey technology, now a multi-billion-dollar industry, owes a great debt to Doc Ewing and the other scientific pioneers who showed the way.

***

One unforeseen consequence of seismic surveys is the impact it has on ocean life. The noise made by seismic surveys affects the behavior of fish, seals, squid, and particularly cetaceans. Since whales and dolphins communicate with each other using sound, noise in the modern ocean has altered their habitat and is probably very disturbing to them. In the last 100 years, the ocean has become a much noisier place because of the cacophony of ship's propellers, seismic surveys, military sonar, and ship's sonars not to mention Jet Skis and wind turbines in coastal areas.[8]

The other impact of the oil industry is of course pollution from oil spills. The worst oil spill disasters occur when oil tankers sink (like the 1989 Exxon Valdez disaster in Alaska), when oil production facilities fail (like the 2010 Deepwater Horizon disaster in the Gulf of Mexico), or when seafloor pipelines rupture. What many people do not realize is that a volume of oil that may be equal to the amount spilled by humans each year is leaking naturally into the ocean from oil deposits.[9] Natural seepage is a gradual, ongoing process, and ecosystems have evolved that use it as a food source. Accidental spills are ecologically damaging because they result in unnaturally high concentrations of oil at a particular site that kills marine life.

Oil enters the ocean from a combination of different sources: land-based sources (urban runoff, coastal refineries); oil transporting and shipping (operational discharges, tanker accidents); offshore oil and gas facilities (operational discharges, accidents, pipeline failures, blowouts); atmospheric fallout; and natural seeps. Estimates range from an average of 470,000 tons to a possible 8.4 million tons per year for the sum of all of these sources. It is generally agreed that the largest single source is the land-based (urban runoff, coastal refineries) input, although there is little agreement on the absolute values for any source terms.

***

The next part of our story is about the waters that fill the oceans. We will explore how ocean currents have evolved over time, what forces drive them, and the role of the oceans in governing the Earth's climate. The composition of the water comprising the oceans has also evolved over time. While many things about the oceans are constantly changing, there is one factor that has remained constant.

The ocean is always moving.

---

[7] Harris et al. (2016).

[8] Williams et al. (2015).

[9] Kvenvolden and Cooper (2003).

Chapter 5
The Ocean in Motion!

"There is a river in the ocean: in the severest drought it never fails, and in the mightiest floods it never overflows; its banks and its bottom are of cold water while its current is of warm; it takes its' rise in the Gulf of Mexico and it empties into the Arctic Seas; this mighty river is the Gulf Stream."
Matthew Fontaine Maury
The Physical Geography of the Sea, 1855

"A ship is safe in harbor, but that is not what ships are built for."
John A. Shedd
Salt from My Attic, 1928

Abstract The ocean is restless. It never stops moving. The famous American oceanographer Matthew Maury described the Gulf Stream ocean current as "a river in the sea," but all the world's rivers combined transport only a tiny fraction of the volume of ocean currents. The great ocean currents regulate the climate by transporting heat around the globe, taking warm water from the equator toward the poles and cool water from the poles toward the equator. Water evaporates from the ocean, falls as rain on the land, and returns to the sea. In this chapter we will learn about the Coriolis effect, one of the most important concepts in oceanography and its consequences for wind and ocean currents. Eddies shed from the ocean currents can reach to great depths and cause deep ocean "storms." We will answer important questions like: What would happen if ice did not float? Why don't icebergs drift in the same direction as the wind blows? What is storm "wave base"? What does a tsunami wave look like in the middle of the ocean? Why are there new beaches forming on the Arctic coast? What has caused Antarctic sea ice "factories" to close down?

Keywords Hydrological cycle · Salinity · Ooid · Limiting nutrient · Matthew Fontaine Maury · *Flying cloud* · Sverdrup · Gulf Stream · Circumpolar Current · East Australia Current · Rhodolith · Hadley cells · *Coriolis* effect · Benthic storm ·

Ekman transport · Fetch · Significant wave height · Storm wave base · Tsunami ·
Great ocean conveyor · Mertz glacier · *Aurora Australis* · Bottom water · Polynya

Today, at some time, you will undoubtedly turn on a faucet for a drink of water.
What you will witness is a small part of a global cycle that begins with evaporation
from the ocean's surface. Sunshine warms the ocean surface to make water vapor
that rises into the atmosphere to form rain clouds. About 78% of rain falls on the
ocean, a brief, somewhat futile version of the cycle lasting only a matter of hours
to days. Rain that falls on land may take a number of different, longer pathways.
Some will flow into lakes or rivers before reaching the sea. Some will be stored in
reservoirs for human use (to fill your water glass). All of the rivers and streams on
Earth combined discharge about 1.2 billion liters per second into the oceans, an
amount that would refill the ocean basins (if they were emptied) in about
34,000 years.

But some water takes a much longer route before reaching the ocean. Rainwater
that seeps into the ground recharges aquifers and becomes groundwater. Here it may
stay trapped in vast subterranean reservoirs for centuries or longer. Groundwater in
the upper Patapsco aquifer underlying the US state of Maryland is over a million
years old.[1] Some water will fall as snow, slowly accumulating into mountain glaciers or the giant, 3 km thick, ice sheets of Antarctica and Greenland; Antarctic ice
is over 400,000 years old at its base. In every case, water that falls on land as rain or
snow eventually is returned to the sea, a journey that takes an average of about
34,000 years, a journey called the "hydrological cycle." Over the 4-billion-year history of the oceans, the average water molecule has made this journey 118,000 times
assuming the hydrological cycle has operated at the same speed since the ocean was
formed.

Water evaporated from the oceans is transported over land by the prevailing
winds: the westerlies blowing between 30° and 60° latitude and the easterlies blowing along the equator and 30° latitude. For example, the moist air evaporated from
the South Atlantic is transported by the easterlies over South America where much
of it falls as rain in the Andes Mountains. This rainwater flows into the Amazon
River that flows back into the South Atlantic. Meanwhile, evaporation from the
North Atlantic forms moist air that is transported westward which falls as rain in the
Rockies, feeding rivers that flow back into the Pacific Ocean. These examples illustrate how some oceans actually export water to other oceans.

Since evaporation removes freshwater and leaves the salt behind, the net result is
that some oceans export more freshwater than they receive. The North Atlantic and
the Mediterranean are net exporters of fresh rainwater, whereas the Arctic and the
Pacific are net importers. Consequently, the Atlantic and the Mediterranean are
saltier than the Pacific Ocean. This raises some obvious questions: Where did the
salt come from in the first place? Why is the ocean salty?

---

[1] USGS – http://www.usgs.gov/newsroom/article.asp?ID=3246#.UKagdqW5dlI

Rain that falls on land washes sediment into rivers containing salts and other minerals that are dissolved into the river water, which eventually empties into the sea (river water is not 100% freshwater – it contains some dissolved salts). Underwater volcanic eruptions also add salts and minerals to the oceans. The transport of salts and minerals by water from the land to the sea is a one-way trip. This is because when water evaporates from the oceans (to make rain clouds), only the water molecules are evaporated, leaving the salts and minerals behind.

Over billions of years, the oceans have reached the concentrations of dissolved salts (seawater is about 3.5% salt) and minerals that we see today. Nearly all naturally occurring minerals are dissolved in seawater, including sodium, calcium, magnesium, potassium, phosphorus, iron, copper, and zinc. There is even gold dissolved in seawater [and yes people have already considered the possibility of mining gold from seawater, although the cost of extraction exceeds the value of any gold obtained in methods tried so far]. But why is the figure 3.5% salt and not 1% or 10% salt?

The salinity of the ocean is balanced by the inflow of minerals and salts from rivers and volcanos and that which is removed by marine life, by chemical processes in the ocean, and by the formation of evaporite deposits. Marine life takes dissolved elements from the ocean needed for making shells and body tissue (carbon, silica, calcium, sodium, potassium, magnesium, and other nutrients). When a marine organism dies or is eaten, its body is exported to the seafloor sediments where the minerals are buried and taken out of the ocean.

Chemical processes can also remove some elements. When the water becomes saturated with calcium carbonate, calcite crystals can precipitate out of solution spontaneously. In some shallow tropical seas (such as the Persian Gulf, the Bahamas, or the Gulf of Papua) where this process regularly occurs, a spheroidal, coated sediment grain known as an ooid is formed by repeated precipitation while the grain is gently rolled along the seabed by waves and currents. Generally less than 2 mm in diameter, each ooid has at its core a fragment of shell, quartz grain, or some other material that has been repeatedly coated with layer upon layer of calcite crystals.

Sodium and chlorine are not as useful to biota as other elements, and so they are left behind in the ocean to slowly accumulate. The ions of biologically useful elements like calcium have a relatively short residence time of around 1 million years; that is to say the average length of time that a calcium ion stays in the ocean is 1 million years. Iron is very useful biologically, and it has a residence time of only around 200 years. In fact, iron is so useful that when it is used up at a location in the ocean, its absence limits the growth of plants and animals – iron is a limiting nutrient in the ocean.

Ions of less useful elements such as sodium and chlorine have long residence times of around 100 million years. The main way that salt (sodium chloride) is extracted from the ocean is by the formation of evaporite deposits, emplaced when lagoons and sometimes larger bodies of water are cut off from the sea and dry out, leaving the salts stranded. The ocean may have been saltier in the geologic past than it currently is because of the thick beds of evaporites deposited during the Mesozoic that are commonly associated with oil fields.[2]

---

[2] Turekian (2001).

The salinity of the ocean today is an equilibrium figure that has been achieved by a balance between competing processes over geologic time. There is no natural law that determines what ocean salinity must be. The ocean contains an average of 3.5% dissolved salts because that is the figure. It's as simple as that.

***

The famous nineteenth century American oceanographer, Matthew Fontaine Maury, is known as the father of oceanography for good reason. While a US Navy lieutenant, an accident left him lame in one leg, so he took a desk job in 1842 as officer in charge of the "Depot of Charts and Instruments" (later named the United States Naval Observatory). From this humble position, he assembled wind and current charts for the entire world, based on observations from hundreds of logbooks recorded by the officers and crew of sailing ships. He also collated the existing 200 or so deep-sea depth soundings from the Atlantic Ocean and was the first person to show that there was an elevated "plateau" or rise located in the center, the first hint of the existence of the mid-ocean ridge.

Maury organized an international conference in 1853 and persuaded hydrographers from all countries to share their data to create the first global synthesis of wind and ocean currents. These charts revolutionized the sailing industry by establishing the safest sea routes where fair winds could be reliably found together with the directions and speeds of the great ocean currents. Using Maury's charts saved sea captains weeks or even months of sailing time along the major shipping routes and probably saved lives as well, since sailors were less often left becalmed on the ocean without sufficient food or drinking water.

In 1854 the clipper ship *Flying Cloud* set a record sailing time of 89 days between New York and San Francisco. Captain Josiah Perkins Creesy had the help of his wife, Eleanor, who acted as navigator on the voyage. Eleanor had read Maury's *Sailing Directions*, and she planned the voyage using his charts. This was at the peak of the California gold rush and prospectors were clamoring for ships to carry them to the west coast of the United States. Sailing around Cape Horn would often take 200 days or more and the record set by *Flying Cloud,* cutting the travel time by more than 50%, made headline news around the world.

Maury's comparison of ocean currents as "rivers in the ocean" flatters the world's rivers, which are actually dwarfed by any one of the major ocean currents. The best estimate is that, added together, the 921 major rivers on Earth transport about 824 million liters (142 million gallons) per second of freshwater into the oceans[3] (the largest is the Amazon which alone discharges 200 million liters per second). If we add in smaller rivers and other runoff, the figure is probably close to 1.2 billion liters per second.[4] That's roughly equal to the flow of 2 billion garden hoses, barely enough water for each of the 7 billion people on Earth to simultaneously take a cold shower! That's if all the freshwater on Earth were diverted for the sole use of humans and humans already use around one quarter of all freshwater

---

[3] Perry et al. (1996).

[4] Dai and Trenberth (2002).

equal to 0.29 billion liters per second[5] (no wonder there is so much concern over conservation of our world's freshwater resources).

Compared with the transport of ocean currents, 1.2 billion liters per second is a mere trickle. Volumes of water transported by ocean currents are so vast that scientists have created a special unit to record them. This special unit, equal to 1 billion liters (264 million gallons) per second, is called a "Sverdrup" named in honor of the famous Norwegian oceanographer Harald Sverdrup. Ocean currents transport volumes of water that is measured in multiples of Sverdrups. The Gulf Stream transports about 35 Sverdrups, and the Kuroshio Current adjacent to Japan transports about 50 Sverdrups, while the Circumpolar Current (the current that flows around Antarctica) transports a massive 100 Sverdrups (100 billion liters per second). In other words, the Circumpolar Current alone transports a volume of water that is 500 times greater than the Amazon and around 80 times greater than the flow of all the rivers in the world combined.

Ocean currents flow in great gyres around every ocean basin. In the North Atlantic, the famous Gulf Stream flows northward along the east coast of the United States and Canada. The current then turns east, becoming the North Atlantic Drift, to bathe Iceland, Norway, and northern Europe in warm water that greatly moderates the climate of that region. The current then turns southward becoming the Canary Current before finally turning westward and becoming the North Equatorial Current.

Along their pathways, the great current gyres absorb, transport, and release heat back to the atmosphere which not only influences the regional climate; it changes the ocean environment bathed by the current. In the southwestern Pacific, for example, the East Australian current flows southward along the coast of eastern Australia, bringing tropical waters from the Great Barrier Reef into the temperate waters of New South Wales. Along its pathway, the current releases heat, and there is a climatic gradient (tropical to temperate) and consequently a biological gradient in the life that characterizes the continental shelf. Coral reefs exist as far south as Brisbane giving way to a fauna dominated by bryozoans and mollusks along the New South Wales coast. At the transition from corals to bryozoans, there exists an interesting zone of life characterized by a curious type of sediment called *rhodoliths*.

The average rhodolith is a roundish, golf ball-sized stone that is made of layers of calcium carbonate that grow from a type of algae, called coralline algae. This type of algae has a distinctive reddish color, and it leaves a hard, limestone crust on the surface of whatever it happens to grow on. The only way that a rhodolith can grow into its distinctive roundish shape is for it to be rolled over every so often so that sunlight can reach the coralline algae growing on all sides. And what is it that makes these rolling stones roll? The East Australia Current of course!

***

How strong does the current need to be for rhodoliths to keep rolling? How big can a rhodolith grow before it tumbles to a halt? Finding the answers to these ques-

---

[5] Hoekstra and Mekonnen (2012).

tions was the goal of a project that I carried out while I was a research fellow at the University of Sydney. I had teamed up with two senior scientists, Prof. Peter Davies and Dr. John Marshall, who at that time worked for the Australian Geological Survey Organization (AGSO; now called Geoscience Australia), and they had secured the project funding from the Australian government in cooperation with the Japanese geological survey.

Our first task was to take some samples of rhodoliths that had been collected offshore from Fraser Island (located at the southern end of the Great Barrier Reef) using the geological research vessel *Rig Seismic*. When coralline alga dies, the rhodoliths turn a dull gray color and have the appearance of lumpy balls of cement. John Marshall and I placed samples of rhodoliths on the bottom of a recirculating flume and measured the speed of current needed to roll them over.

The next step was to measure the strength of the East Australia Current in the area where the rhodoliths had been collected. The idea was to deploy an array of current meters in a line offshore from Fraser Island to the edge of the continental shelf in about 140 m water depth. The current meters were provided by the University of Sydney, and we hired a boat from the University of Queensland in Brisbane to deploy them. As I was the junior team member, it was my job to carry out the current meter deployments.

Early in the summer of 1992 (that's in November and December down under), we set sail for the first of our deployments. The boat we had hired from the University of Queensland was a 50 foot fishing trawler called the *Sea Wanderer*. Used mainly for biological sampling in the relatively protected waters of Moreton Bay, our project was taking the Sea Wanderer into conditions that were going to test the limits of its sea-going capability. Sea Wanderer had stabilizers that hung on chains deployed on either side of the vessel from booms that could be lowered outboard to stop the boat from rolling side to side in heavy seas.

My co-worker from the University of Sydney was Dave Mitchell, an experienced sailor and electronics technician. Dave and I had worked together for several years in Torres Strait and off the coast of Sydney on various projects. Our current meters were to be deployed using a U-shaped mooring. In this design, the current meter is held 1 m above the seabed by a float held in place by a 100 kg (220 pound) iron weight. The weight is attached to a stainless steel ground wire which runs to a second iron weight, and from that a heavy polypropylene rope is attached to a plastic float that is at the sea surface. Attached to the surface float is a string of four or five Styrofoam pellet balls tethered by a rope. The pellet string is how you recover the current meters, by throwing a grappling hook and pulling them aboard. The advantage of the U-shape mooring is that if the surface floats are cut loose or sunk by a passing ship, you can fish for the ground wire using a grappling line to recover the valuable current meter.

To deploy the current meters, you start by lifting the first iron weight off the deck by the ground line using the ships winch. The current meter and its float are man-handled overboard, and the ground wire is paid out from the winch to lower the instrument carefully to the seabed. The ground wire must be longer than the water is deep in order for the weight to reach the seabed and give you some slack. Then the wire is quickly pulled off the winch and the second iron weight is lifted off the deck, this time using the heavy polypropylene rope. The second weight is then lowered to the seabed, and once it is on the bottom, the excess rope is pulled off the winch. The surface floats are then quickly attached, and the pellet floats are cast overboard.

This sounds easy in theory, but it involves a couple of tricky bits. First is that the captain has to keep the boat on top of the deployment site as closely as possible. This meant that the *Sea Wanderer* had to back into the East Australia Current since the current meters were deployed from the back of the boat. Each time one of the iron weights was on the bottom, the excess wire and rope must be quickly taken off the winch. If the boat drifted down current, then the full weight of the weight came onto the wire or rope and made them impossible to remove.

The first current meter we deployed was in 35 m water depth, close to the Fraser Island coast where the East Australia Current is weakest. With each successive deployment in deeper water, further from the coast, the current grew stronger and the captain had greater difficulty keeping the boat on location. At the most seaward station in 140 m depth, we had to work quickly to pay out the wire and deploy the current meter. Here the East Australia current was running swiftly at around three knots and *Sea Wanderer* struggled to stay on station. A brisk northeasterly wind made the seas choppy and waves broke over the stern of the vessel as we fastened on the surface floats and tossed the pellet floats over the stern.

The events of that last deployment are clearly etched in my memory.

As the *Sea Wanderer* turned away from the mooring, the pellet floats caught on the stabilizer on the port side. The rope became tangled with the stabilizer fin and as the current carried us along the line grew taunt. Our boat was effectively anchored to the seabed and the East Australia Current began to drag the boat around sideways. The boom that the stabilizer was attached to acted like a lever to pull the boat over and *Sea Wanderer* listed dangerously, lifting the propeller out of the water. Waves broke over the deck and for a few seconds I think we came very close to capsizing. I remember Dave Mitchell calmly ducking down below and returning on deck to hand me a life jacket!

Luckily the pellet line rope broke and *Sea Wanderer* bobbed back into its proper, upright orientation, escaping any serious damage. Learning from this lesson, we hired a much larger boat after that to complete the project, and we did eventually recover all the current meters.

The meters showed that over a period of 1 month, the East Australia Current flows at speeds of up to 1.30 m/s (2.5 knots) measured close to the seabed in 80 m water depth and nearly twice that speed near the surface. That is strong enough to roll 5 cm diameter rhodoliths along the seabed.[6] The largest rhodoliths we found were over 7 cm (three and a half inches) in diameter. Ours was a small triumph, but it gave us the information we needed, and it was the first time anyone had measured such strong currents in that region.

***

----

[6] Harris et al. (1996).

Every ocean basin has its own currents. They have been travelled by sailors over the centuries, and ocean life has evolved to utilize and depend upon them to fulfil their life cycles. Each current has its own name, different properties, and special temperaments, but they all have one thing in common: they all flow in gigantic circles around the edges of the ocean basins.

Why do currents flow in circles? What forces drive the great ocean currents?

If the Earth were completely covered by ocean, without any large continents, the ocean currents would default to a series of latitude-parallel streams encircling the globe. "Why is the default for ocean currents a series of latitude-parallel streams encircling the globe" I hear you ask. The answer has to do with Matthew Maury's wind charts and the transfer of heat over the surface of the Earth, because the ocean currents are driven by wind.

Heating of the atmosphere at the equator and cooling at the poles creates a global heat imbalance; to achieve a balance, warm air must move toward the poles and cold air toward the equator. Heating at the equator causes warm air to rise; it is cooled high in the atmosphere before it descends at a latitude of around 30° (both north and south of the equator; Fig. 5.1a).

At the poles, the cold (dense) air sinks and flows toward the equator, becoming relatively warm air at 60° latitude. Here, it rises before sinking again at the poles. So, altogether there are three bands of moving air in each hemisphere, with zones of rising warm air at the equator and at 60° latitude (both north and south) and zones of sinking cold air at 30° latitude (both north and south) and at both of the poles. These six bands of moving air (three in each hemisphere) are known as Hadley cells, named for George Hadley, a seventeenth-century English meteorologist. The physics of Hadley cells is universal; it applies on Earth and also provides an explanation for the colored bands of Jupiter's atmosphere, for example. Hadley cells have existed on the Earth from the time it had its first atmosphere.

Wind blowing across the ocean surface is deflected by Earth's rotation (the *Coriolis* effect) which gives rise to the northeast and southeast trade winds on either side of the equator and the westerlies between latitudes of 30° and 60°. Since the Earth rotates once per day, at the equator the Earth is spinning at a speed of around 1600 km/hour, while at the poles, the angular speed of rotation is zero. This means any object moving across latitudes over the Earth's surface appears to be deflected to the right in the northern hemisphere and to the left in the southern hemisphere.

Textbooks often depict a ballerina on one of the poles, showing how she rotates once a day as the Earth spins. If she were to be moved in any direction from the pole, she would appear to be deflected, because she retains that spin (angular momentum) acquired at the pole. The spin of the ballerina is greatest at the pole, but it decreases to zero at the equator. This is because at the pole, she is standing parallel to the axis of Earth's rotation, but at the equator, she is at exactly 90° to the axis, and so she has none of the Earth's spin.

These two aspects of the *Coriolis* effect, decreasing angular speed with latitude from equator to the poles and decreasing spin from the poles to the equator, result in something special happening at the equator. Here, an object that is set in motion *along* the equator (i.e., parallel to the equator) is immune from the *Coriolis* effect.

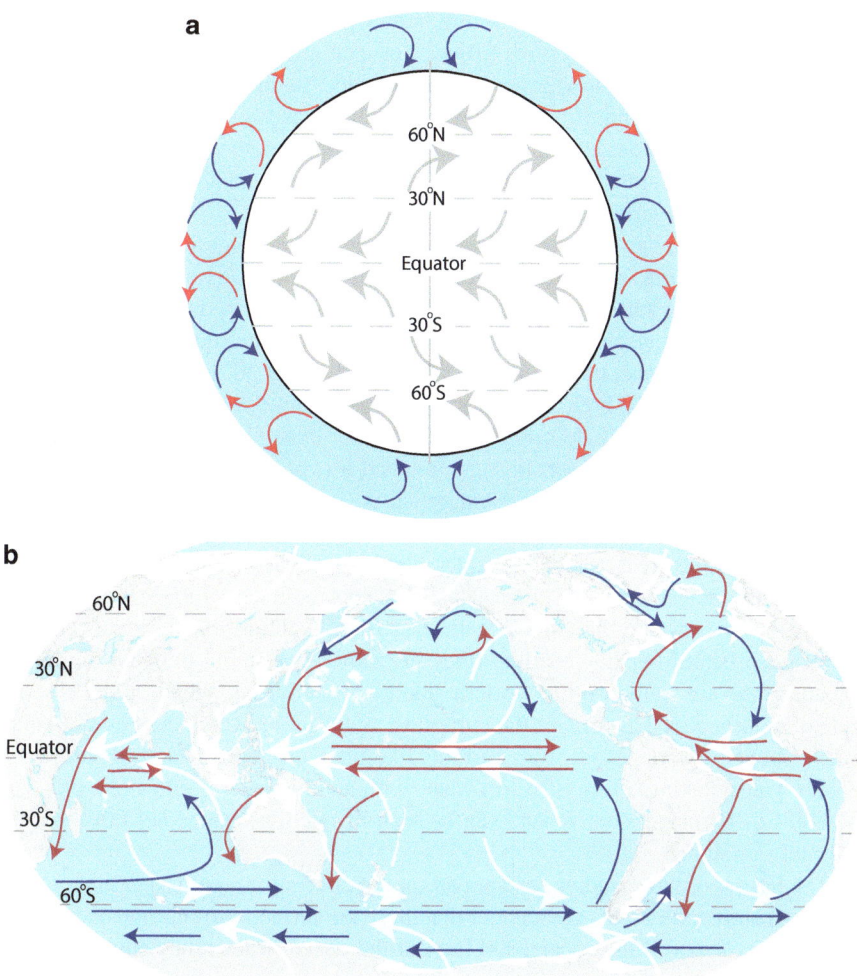

Fig. 5.1 (**a**) Sketch showing Hadley cells in the atmosphere in cross-section (not to scale). Red arrows represent warm, moist, rising air and blue arrows represent cool, dry, descending air. (**b**) Map showing surface ocean wind and currents in the world ocean. Surface winds (gray arrows) are shown diverging from 30° latitude (north and south), being deflected by the Coriolis force and converging at the equator. Winds also converge at 60° latitude (north and south) merging with air descending at the poles. The winds drive the main ocean currents, both warmwater (red arrows) and cold-water (blue arrows) currents

Ocean currents are driven by the trade winds and westerlies, and they flow around the globe in an east-west or west-east, latitude-parallel, direction. These currents are blocked by continents, which stop them from flowing around the globe in a continuous loop. Instead they are deflected by land to form the great ocean gyres that flow around the major ocean basins (Fig. 5.1b). The only exception is the Circumpolar Current which flows unhindered, in a giant circle around Antarctica.

Strong winds blow between the latitudes of rising and descending air. The "roaring forties" and the "furious fifties," for example, are sailor's terms for the strong trade winds that blow around these latitudes. In contrast are the calm wind (doldrums) areas around 30°, also known by sailors as the "horse latitudes." The origin of that name has several possible explanations, one involving Spanish sailing ships being becalmed, without drinking water, and the sailors having to throw their dead horses overboard.

Hadley cells explain two other of Earth's climatic phenomenon: deserts and rain forests. Rising (warm) air at the equator and at 60° latitude carries a lot of moisture. As it rises, the moisture condenses and falls as rain. This is one reason why the tropical and temperate rain forests are typically found at the equator and at around 60° latitude. In contrast, the air sinking at the poles and at 30° has had the moisture removed as it travelled high in the atmosphere, making it very dry. This explains why there are commonly deserts found at 30°, for example, the Sahara Desert in Africa and the Simpson Desert in Australia, and it explains the very low precipitation found to occur at the poles (Antarctica is, in fact, the driest of the Earth's continents).

Currents of the great ocean gyres interact with the coastline to spawn eddies that spin off into the center of the gyres. This happens where any of the strong western boundary currents like the Gulf Stream, Kuroshio, Agulhas, or other currents are deflected offshore by the coastline (they are called "western" boundary currents because they are found on the western side of ocean basins, which is the eastern side of continents). This causes the current to make a loop which creates a circular eddy. Loops of warm water from currents like the Gulf Stream capture parcels of colder water making a ring of warm water with a cold center. Eddies like this can be 100's of kilometers across and can be seen in satellite images because of their thermal signature.

The largest of these eddies can spin around in circles for months at a time, while they drift across the ocean. Remember that these are surface currents and occur only in the upper few 100 m (upper 1000 feet) of the ocean. But as they spin, the eddy will gradually extend deeper and deeper into the water column until they eventually touch down on the ocean floor. When that happens, there is a storm on the seabed!

***

The deep ocean is not a stagnant stillness, but a dynamic place with its own "weather." The realization that there are "benthic storms" in the deep sea is a relatively new idea from the early 1980s. Marine geologists were puzzled by samples of seawater collected in the 1960s and 1970s from many locations near the ocean floor found to contain sediment in suspension, presumably eroded from the seabed. What could be the cause of the so-called *nepheloid* layer of suspended sediment (from the Greek nephos, meaning "cloud")? The solution came from bottom photographs showing ripples, furrows, and grooves made by strong bottom currents and by current meters anchored close to the seabed.

Proof of the existence of storms effecting the deep ocean floor was provided by the High Energy Benthic Boundary Layer Experiment (HEBBLE).[7] The experiment

---

[7] Nowell et al. (1985).

ran for several years at a site in 4830 m water depth in the North Atlantic, collecting measurements of currents near the seabed, water samples to measure the amount of sediment suspended in the water column and underwater photographs of the seafloor. The results of HEBBLE were astonishing and transformed our thinking of the deep ocean environment.

We know now that benthic storms can last from 2 to 22 days and are mostly from 3 to 5 days in duration. At any given site, there can be as many and 8–10 storms per year. Near-bed currents during benthic storms attain speeds of up to 0.73 m/s – this may not sound like much but when you take into account the density of water compared with air, a near-bed current speed of 0.73 m/s is comparable to a 32 m/s (70 miles/hour or 11 on the Beaufort scale), which is a violent windstorm!

What affect would such storms have on animals living here? They must be adapted to survive in such an environment, where episodic disturbances like benthic storms add a bit of spice to an otherwise quiet existence. Benthic storms would transport sediment, microorganisms, larvae, and juveniles long distances along the storm's path. Sediment clouds may extend over distances of thousands of kilometers[8] from the storms epicenter.

We know now that benthic storms occur in all oceans and at depths of over 5000 m. They are formed where a large eddy persists in the ocean, causing the water column to spin into deeper and deeper water, finally touching down onto the seabed, like a tornado or cyclone on land. Although first discovered and studied in the North Atlantic, computer modelling studies suggest that benthic storms affect around 20% of the ocean floor and are most common in areas next to the major ocean currents, especially the Circumpolar Current around Antarctica.

Global climate change driven by human emissions of greenhouse gases has warmed the surface ocean and also caused a global increase in wind speed. The faster winds are driving the western boundary currents faster and making them wider and warmer. The western boundary currents are warming twice as fast as the rest of the ocean.[9] Does this mean that benthic storms are growing in intensity and becoming more frequent?

Surface currents like the Gulf Stream, flowing in gyres around the ocean basins, are only a part of the ocean circulation system because, of course, ocean circulation is three-dimensional. Surface currents like the Gulf Stream flow in the upper 300 m (1000 feet) or so of the ocean; only the eddies they shed reach to the deep abyssal seabed. But there are other, deepwater currents flowing near the ocean floor just as there are currents flowing at the sea surface. These deep ocean currents are obviously not driven by the wind, so what causes them to occur? What processes drive the deep circulation of the oceans?

The answer is related to the fact that ice floats.

***

Water is really quite an amazing substance. Not just because it makes life possible as we know it. But water is also amazing for its physical properties, one of which is what happens when it changes from a liquid to a solid state.

---

[8] Hollister and McCave (1984).
[9] Miller (2017).

Most substances attain a peak density in their solid (crystalline) phase. Molten (liquid) iron, for example, has a density of 6.98 g per cubic centimeter (g/cm³) but reaches 7.87 g/cm³ as a solid. This means that when molten iron cools, the crystals of solid iron will sink into the liquid iron. As a pot of molten iron cools, we would find the solid iron on the bottom of the pot while the liquid remains on the surface.

Water behaves in the opposite way.

The density of distilled water at 20 °C is 0.9982 g/cm³. The density increases as it cools down and reaches a peak of exactly 1 g/cm³ at a temperature of 4 °C (in fact, the density of water is the basis for the metric system, which relates the densities of all substances to water). Between 4 °C and 0 °C the density of water decreases again until it freezes at 0 °C. Ice has a density of 0.915 g/cm³, significantly less dense than water. And so, ice floats on the top of the liquid water.

The density maximum attained by water at 4 °C could theoretically lead to some strange depth profiles of temperature in freshwater bodies that are stratified by temperature. Under most conditions, a freshwater lake would have the warmest waters at the surface, and coldest (denser) water would be found at the bottom of the lake. But in a very cold freshwater lake, it is theoretically possible that the temperature of the water at the bottom of the lake could be close to 4 °C, while at the surface, the temperature is close to zero. In other words, the warmest water is at the bottom of the lake instead of at the surface as one might expect.

Seawater is slightly denser than freshwater because of the dissolved salts it contains. The saltiness, or salinity of seawater, is typically around 35 parts per thousand (35 ppt equal to 3.5%) by mass. The density of seawater increases with salt content. Seawater with a salinity of 35 ppt at 20 °C is 1.0248 g/cm³, and with a salinity of 40 ppt at 20 °C, the density increases to 1.0286 g/cm³. This difference in density seems small, but it is enough to cause water bodies to sink or rise, always seeking to establish an equilibrium of the densest (cold and salty) waters at the bottom with warmer and fresher water at the surface.

Seawater density also increases with decreasing temperature, but unlike freshwater it does not reach a density maximum before it freezes. It just keeps getting more and more dense as it cools. At a temperature of 0 °C, seawater with a salinity of 35 ppt has a density of 1.0281 g/cm³ and at −2 °C density of 1.0282 g/cm³. Seawater freezes at around −2 °C at which point it crystalizes forming ice. Sea ice is made only of freshwater (the salt is rejected as seawater freezes), so it still has a density of 0.915 g/cm³. And so, (sea) ice also floats on top of seawater.

What would happen to the world if sea ice did not float?

Ice forms every winter in the polar latitudes, and if ice did not float, it would sink to the bottom. Ice on the seabed would melt slowly if at all in summer because warm water is less dense and it stays at the surface while the coldest seawater would remain at the bottom of the ocean, next to the ice. The Arctic Ocean and the Southern Ocean basins would therefore gradually fill with ice that formed each winter. The flow of warm equatorial water into the polar regions is what keeps those water bodies above zero in temperature. But the warmer waters flow into the polar regions at the surface; cold water stays at the bottom.

As the basins filled with solid ice, the flow of water would be blocked, and so the frozen ocean would gradually expand into the temperate oceans. Ice forming in the winter months would sink to the bottom even at temperate latitudes and eventually fill those basins as well. The thermal mass of ice would cool the Earth's temperature further until, eventually, all of the ocean basins would be frozen solid. This would lead to a snowball Earth scenario from which the Earth could never escape. Any life that existed on Earth would have to cope with sub-zero climate conditions. In short, the fact that ice floats makes life as we know it possible on Earth.

The behavior of floating icebergs revealed one of the ocean's mysteries to the famous Norwegian oceanographer and explorer, Fridtjof Nansen. During his expedition to the Arctic aboard the *Fram* in 1893–1896, he noted that icebergs seemed to drift in a direction at an angle to the wind direction (not exactly in the same direction that the wind was blowing). Why did that happen, Nansen pondered? A young Swedish crew member and student of Nansen's, Vagn Walfrid Ekman agreed to tackle the problem after the *Fram* returned from its voyage. His PhD thesis describes the phenomenon now known as *Ekman transport*, and the theory and mathematical formulae were described in a scientific paper published in 1905.[10]

When wind blows over a broad (10's of km long) expanse of the ocean, Ekman reasoned that the surface friction drags the upper layer of water along with the wind. The upper layer of water, in turn, drags along water in the next layer down and so on through the water column. The speed of the surface layer is the fastest, and friction causes each layer below to move slower and slower until at some depth there is no movement. But the top layer is deflected slightly by the *Coriolis* effect. Each layer is deflected a bit more, to the right-hand side in the northern hemisphere and to the left-hand side in the southern.

When Ekman added up all the layers, he found that the average movement of the whole water column is actually at 90° to the wind direction (90° to the right-hand side in the northern hemisphere and 90° to the left-hand side in the southern). This is known today as *Ekman transport*, and it is a fundamental concept of oceanography. The direction taken by an iceberg depends on how deep it extends below the surface and thus how many layers of moving water it passes through; if it has a shallow keel, it will move along with the corresponding (uppermost) water layers at an angle of much less than 90° (and the freeboard of an iceberg acts like a sail also pushing it along in the direction of the wind). A deeper keel makes the iceberg move at a greater angle to the wind direction.

Icebergs aside, Ekman transport explains the origin of wind-driven currents in the ocean. It is a universal theory that seems to fly in the face of common sense; wind drives a current at an angle of 90° to the direction that the wind blows (Fig. 5.2). How can this be? It is one of the consequences of living on a rotating planet and is the source of great consternation for first-year undergraduate oceanography students.

The application of Ekman's theory requires that certain rules of scale are followed. The wind must be blowing over a sufficient length of ocean to drive a current

---

[10] Ekman (1905).

Fig. 5.2 Diagram illustrating the Ekman spiral and wind-driven transport, whereby the surface current moves in a direction at 90° to the direction the wind blows

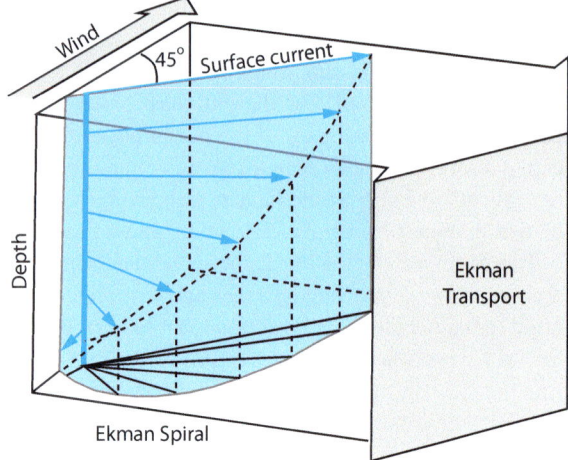

at 90°, typically several tens of kilometers. The water depth must also be great enough for an Ekman spiral to develop, nominally greater than 65 m (130 feet). Ekman's theory applies on the ocean; in shallow water the current will flow more or less in the direction the wind blows.

The practical value of the discovery of Ekman transport is enormous. Ekman's equations are incorporated into every calculation of wind and ocean currents used for applications ranging from shipping, pollution dispersal to fisheries. It was truly a giant leap forward for the science of oceanography.

There is one other thing that is created when wind blows across the surface of the ocean: waves.

***

The ocean is a dangerous place – on average two large ships are lost per week out of the global fleet of around 85,000. Around 2000 sailors are lost at sea each year. The main reason is stormy seas when things go badly wrong. For example, storm waves may cause the cargo to shift causing a ship to capsize. Ships also collide, run aground, and catch on fire. I am talking only about expensive ships that are insured for which statistics are available.[11] Nobody knows how many small (unregistered and uninsured) fishing vessels are lost at sea every year.

Waves washing over the bow of a ship or crashing on the shore are common images depicted in artist's impressions of what constitutes the character of the restless ocean. Perhaps because of this restless motion, people have endowed the ocean with personality, as if it were a living thing. The ocean is often described using any number of poetic adjectives. It is treacherous, angry, dangerous, hostile, and merciless, but it can also be tranquil, calm, gentle, embracing, and caressing. It is benign, timeless, and ageless but also mysterious and beyond comprehension.

---

[11] https://maritimeintelligence.informa.com/products-and-services/lloyds-list-intelligence/casualty-reporting

The ocean may be indifferent to human beliefs and perceptions, but nevertheless it has played a major role in defining different cultures. Since the beginning of human civilization, humans have lived and worked on or beside the ocean. Many cultures created gods of the sea, to whom they worshipped to pray for bountiful catches and safe passage. The Māori people of New Zealand worshipped Tangaroa, the god of the sea. Neptune is the god of the sea in Roman religion, while the ancient Greeks worshipped Poseidon.

Swell waves are created by wind shear over the ocean surface. The important variables are the wind speed, the length of time the wind blows at that speed (duration), and the distance over which the wind blows (fetch). Strong winds blowing for a long time over a long fetch make the largest waves.

All waves are described by three measures: wavelength (distance between wave crests), wave height (vertical height from peak to trough), and wave period (length of time between successive wave crests). The exact location where the largest waves in the world's oceans are found varies from year to year, depending on the occurrence of storms and prolonged high wind speeds over a large fetch. The westerly winds blowing over the Southern Ocean are uninterrupted by any land, meaning the fetch is infinite since wind can blow around the globe without stopping between 50° and 60° latitude. However, waves in the North Atlantic are measured to attain an equal height to those in the Southern Ocean. This is because the most extreme wave heights are created by the strongest winds blowing for the most length of time over a certain distance (fetch). It is the wind strength and length of time and not the fetch that are limiting factors. The fetch is large enough in all the oceans for the largest waves to be created.

A statistic that is often used to describe waves is the "significant wave height," which is the average of the highest one-third of waves. Using radar transmitted from satellites, oceanographers have mapped the significant wave height of swell waves for the last few decades. As expected, the higher latitudes during winter months experience the strongest winds (roaring 40s and furious 50s) and the largest swell waves. Significant wave heights in winter are commonly 6 m (19 feet) in the Southern Ocean, the North Pacific, and North Atlantic. Based on an analysis of satellite data spanning more than 20 years, oceanographers have detected a global trend for increasing wind speed and significant wave height, attributed to climate change.[12]

The effects of waves as they propagate across the ocean are most apparent at the sea surface, but once the wave moves into shallow water, the effects start to be felt on the seabed. This occurs when the depth of water is about half of the wavelength. That is, waves that are 100 m (300 feet) apart begin to stir sediments on the seabed in water that is 50 m (150 feet) deep. Larger waves that have longer periods and greater wavelengths can stir seabed sediments at greater depths. Modelling of waves and the resultant sediment motions shows that the largest waves in the North Atlantic are able to mobilize fine sand at depths of up to 234 m (768 feet) on the Rockall Bank.[13]

---

[12]Young et al. (2011).

[13]Harris and Coleman (1998).

Thinking about the maximum depth at which waves can mobilize sediment points to another perspective – the depth below which the largest storm waves are unable to mobilize sediment. This depth is referred to as storm "wave base."

Waves of different periods traversing the same space will interfere with each other in a destructive way (cancelling each other) or in an additive way, making even larger waves. In extreme cases, groups of waves may merge to create a giant "rogue" wave having heights attaining as much as 60 m (190 feet) amplitude. Rogue waves can be ship killers – they are attributed to the loss of seven large "supercarrier" types of cargo ships per decade.[14]

You may have seen the famous photograph of a huge wave crashing over and around the La Jument Lighthouse off the coast of France, while the lighthouse keeper looks out of the lighthouse door. The photo was taken from a helicopter during a storm event in December 1989 capturing the image of a rogue wave engulfing the lighthouse structure.

Tsunami are a different sort of wave altogether. They are triggered by an instantaneous action that causes a massive displacement of water: an underwater earthquake or landslide, a meteorite impact, or a volcanic eruption, for example. Formerly known as "tidal waves," the name was changed because these waves have nothing to do with tides caused by gravitational interactions between the oceans, the moon, and the sun. To avoid confusion, experts have chosen to use the Japanese word "tsunami" which translates to "harbor wave."

Japan is prone to experience regular tsunami because Japan is located adjacent to a deep ocean trench with its subduction zone and the archipelago is regularly jostled by seismic events. Tsunami are deceptively powerful in nature. Their wavelengths are measured in hundreds of kilometers, and they travel at hundreds of kilometers per hour, but out on the deep ocean, their amplitude is only around 1 m. Their height on the open ocean is too small to be detected using satellite radar systems. Japanese fishermen working out at sea would hardly notice the passage of a tsunami lifting their boats slowly up and down by 1 m over a period of 30 minutes or so. But when they returned home, to their harbors, the devastation caused by a tsunami would be all too obvious. Of course, tsunami can occur anywhere along the coast, not just in harbors, so we use the term "tsunami" with this broader meaning.

Suffice it to say that tsunami are very big waves. The Boxing Day tsunami of December 26, 2004, was triggered by a magnitude 9.2 earthquake in the Indian Ocean. The "megathrust" earthquake was triggered by a massive 15 m (50 foot) vertical displacement of seafloor along 1600 km (1000 miles) where the Indian plate is being subducted under the Burma Plate of Sumatra, Indonesia. The resulting tsunami impacted coastal communities from Java to Sri Lanka with waves up to 30 m (100 feet) high; it killed between 230,000 and 280,000 people in 14 countries and was one of the deadliest natural disasters in human history.

The 2011 Tōhoku tsunami on Japan's Pacific coast was triggered by a 9.1 magnitude earthquake, where the Pacific Plate is being subducted under Japan. The

---

[14]Dysthe et al. (2008).

wave was estimated to have attained a height of 38.9 m (128 feet). The disaster killed nearly 16,000 people and caused billions of dollars in damage. A meltdown at the Fukushima nuclear power plant was caused by damage to back up generators from tsunami-induced flooding.

The most common waves in the ocean are wind-generated, surface swell waves. Satellite altimetry data shows that the largest ocean swell waves are found at high latitudes, but not in the polar seas. This is because of sea ice. Once the sea surface has frozen over, the wind shear loses its effect. Consequently, the polar seas are relatively calm in spite of the strong winds that blow across them.

Until recently that is.

Waves in the Arctic are getting larger and more destructive every year. The reason is because the sea ice is melting due to anthropogenic climate change. Arctic sea ice extent shows a steady decline from the date of accurate satellite monitoring (1979) to the present, from around 7.5 million km^2 in 1979 to around 4 million km^2 in 2016 (summer sea ice minimum, related to ice extent measured in September). This is a reduction of nearly 40% in area. Consequently there is open water over large parts of the Arctic Ocean that were previously ice-covered year-round. And it means that swell waves are forming in the Arctic during summer months for the first time in human history. Areas of the Arctic coast that have never previously been impacted by waves are eroding and forming beaches. Thawing of the permafrost is a key driver for retreat of Arctic coasts (by as much as 30 m/year), but wave attack of formerly sea ice-protected coasts is also a factor.[15]

Sea ice not only dampens the swell waves in polar seas; it plays another critical role in ocean circulation. When seawater freezes, salt is rejected by the ice crystals. It makes freshwater ice, leaving behind very salty brine droplets. The brine seeps back into the ocean surface making it slightly saltier. Basically, freezing seawater leaves freshwater ice floating on cold, salty, and therefore dense seawater. The dense water sinks toward the seafloor.

And this simple process, the rejection of salt by frozen seawater, is the engine room that drives the deep ocean circulation.

***

Deep ocean circulation and the flow of water near the deep ocean floor are among the least well-understood processes of the oceans. The main reason is that the underlying technology for making observations of water properties from a ship has not advanced much since the days of the *Challenger*. Essentially, the main way to measure the deep ocean is still to lower a probe or sampling device over the side of a ship, paying out wire until the desired depth has been reached. Our probes and sampling devices have evolved considerably since the 1870s, but the basic idea is the same, and oceanographers still have to spend many hours watching the great wheels of shipboard winches paying out and reeling in thousands of meters of wire rope. The deeper the water, the more wire you must pay out and reel back in and the longer it takes to complete a station.

---

[15] Overeem et al. (2011).

It is not so much that it is just plain boring to keep a research ship sitting on station for hours on end while winches reel wire out and in; it is also very expensive; oceanographic research vessels cost upward of US $50,000 per day to operate. Consequently, only 6% of hydrographic observations have been collected from below 2000 m, even though 51% of the ocean's volume lies below this depth.[16] The lack of data on deep ocean water properties is mirrored in ocean computer models, which have their highest resolution at the surface and very poor resolution in the deep ocean.

Observations of the ocean's temperature and salinity are also being taken these days using robots, known as Argo floats. The name "Argo" is from Greek mythology; it was the name of the ship used by Jason and his Argonauts in the search for the golden fleece. Today there is a fleet of around 3800 Argo floats, operated by an international science consortium of 30 countries.[17]

Argo floats are deployed from research ships and left to drift around the ocean gyres. They are programmed to sink to a depth of 2000 m before rising to the surface, taking measurements of temperature and salinity along the way. The data are radioed back to land via satellite when the float makes a brief stop at the surface. This technology obviously does not work where there is sea ice present since Argo floats must reach the surface in order to radio back the information. This means that Argo floats cannot be deployed in the Arctic or close to Antarctica, although there have been trials of devices that can be left on floating sea ice floes with radio transmitters.

A new Argo float designed to descend to 4000 m depth was tested in 2016 and will start to be deployed more broadly in the future. Meanwhile, Argo floats are collecting thousands of observations every day in the upper 2000 m of the ocean to help us understand ocean circulation, weather forecasting, as well as tracking the impacts of global climate change. Results show that in the global surface ocean (upper 700 m), the average temperature has warmed by 0.168 °C (0.302 °F) in the last 50 years.[18]

Satellites also measure the temperature of the ocean surface water. The US National Oceanic and Atmospheric Administration (NOAA) collects a daily image of the global ocean at a resolution of one-fourth of a degree (15 nautical miles) based on measurements of infrared and microwave radiation. But the properties of the deep ocean waters below 2000 m and their movements are poorly understood.

The coldest, saltiest, densest, so-called bottom water is formed in the polar seas and sinks to the bottom of the ocean where it flows toward the equator. Along the margin of Antarctica, the formation of sea ice drives the production of bottom water. The engine rooms for Antarctic sea ice production are locations where strong offshore winds drive the ice away from land, making open, ice-free areas known as

---

[16] de Lavergne et al. (2016).

[17] http://www-argo.ucsd.edu

[18] Levitus et al. (2009).

"polynyas" (the Russian word for "lake"[19]). Seawater freezes making sea ice; the salt is rejected and added to the frigid (−1.8 °C) surface waters, which then becomes very dense.

Antarctic wind blows offshore at hurricane force for good reason. Cold air sinks and flows downhill over the Antarctic ice sheet from an elevation of over 3000 m in the center of the continent, down to sea level at the coast. These strong offshore winds are called "katabatics" because they are mostly gravity-driven; they blow the newly formed sea ice offshore, keeping the polynyas open, and the process continues. Polynyas are located all along the Antarctic margin, and these "sea ice factories" run all winter long, forming dense bottom water that sinks into the Southern Ocean.

The cold, salty, dense water that is created in polynyas is called Antarctic Bottom Water. From its origins on the continental shelf, Antarctic Bottom Water flows down the continental slope and sinks to the abyssal depths of the Southern Ocean. It flows downhill under the influence of gravity, filling seafloor depressions which then overflow. Very slowly, at speeds of a few centimeters per second, it makes its way into the Indian, Pacific, and South Atlantic basins. Antarctic Bottom Water accounts for around 30–40% of the mass of the ocean, and it covers two-thirds of the deep ocean seafloor. It is thought that around 17 Sverdrups (17 million cubic meters per second) of bottom water are produced in the Antarctic, but the exact figure is uncertain.[20]

In the North Atlantic Ocean, bottom water is created where the Gulf Stream splits into two parts off the coast of Great Britain. One branch warms the seas around Great Britain on its way to Norway, making British weather just barely tolerable ("weather" is undoubtedly the favorite topic of conversation of British people). This branch becomes the northward flowing Norwegian current, and the climate of Norway would be much colder without its warming effect. This warm current also inhibits sea ice forming along the Norwegian coast which is most critical. This is because sea ice acts like a blanket, trapping heat in the ocean. The climate of Norway would be much colder if sea ice formed in the Norwegian Sea.

A second branch of the Gulf Stream splits off to the northeast and makes a detour around the southern part of Iceland before dissipating along the coast of Greenland. Along the way, both branches of warm and salty Gulf Stream waters are transformed into cold and salty bottom water. The Norwegian Current cools and sinks around the island of Svalbard, whereas the eastern branch cools and sinks off the coast of Greenland.

This process leads to one unfortunate environmental consequence. Plastic debris floating on the surface of the Gulf Stream is carried into the Arctic on a one-way trip. Bottom water exits the Arctic at depth, leaving the plastic behind, making the Arctic a net importer of North Atlantic garbage.

Meanwhile, in the Arctic Ocean, sea ice is formed all winter long, making cold, salty bottom water. This water is mainly trapped in the Arctic Ocean basin, which is

---

[19] Massom et al. (1998).
[20] Johnson (2008).

closed on all sides apart from two passages: one is the Denmark Strait located between Greenland and Iceland. The other is a bathymetric feature called the Faeroe Bank Channel located between Iceland and the Faeroe Islands, located off the northern tip of Scotland (the Bering Strait is far too shallow for bottom water to escape through it into the North Pacific).

The sill of the Denmark Strait is only 700 m (2300 feet) deep, and the dense bottom water flows over the sill and cascades down a 3.5 km (2.2 mile) tall cliff, into the depths of the Irminger Sea (located between southeast Greenland and Iceland). This is one of the greatest "waterfalls" in the ocean. The largest waterfall on land is Angel Falls in Venezuela, which is 3212 feet (979 m) tall. But compared with waterfalls in the deep sea, Angel falls is merely average in size. Arctic waterfalls are the origins of North Atlantic Deep Water, which descends from the two Icelandic passages to depths of up to 4000 m.

North Atlantic Deep Water flows southward the length of the North Atlantic, at a rate of about 17 Sverdrups. Some North Atlantic Deep Water reaches the Southern Ocean. But Antarctic Bottom Water is colder and saltier, so it sinks beneath the North Atlantic Deep Water. Overall, North Atlantic Deep Water covers around one-third of the deep ocean seafloor.[20]

The volumes of sinking water in the polar seas are balanced by water rising from the depths to the surface in different locations. These are called "upwelling" zones, and they are found in the North Pacific, the northern Indian Ocean, and the Southern Ocean, among other places (more on this later). The continuous loop of water sinking in the poles, flowing around the deep ocean basins before finally rising to the surface, is known as the great ocean conveyor[21] (Fig. 5.3).

The conveyor turns slowly. It takes around 500 years for North Atlantic Deep Water and around 900 years for Antarctic Bottom Water to complete one circuit. The conveyor transports cold water from the poles to the equator to balance the heat transported in the warm surface currents that flow from the equator to the poles. The proof that there is a great ocean conveyor can only be obtained from research ships equipped with sensitive instruments capable of collecting samples of seawater from the greatest ocean depths. And to collect samples from the sea ice zone, you need a special kind of ship, an icebreaker, in order to access these hostile environments. In all such endeavors, there is an element of danger, even with all the modern technology we have today. Things can go wrong at sea. And the sea is as unforgiving today as it was for any ancient mariner.

***

On July 15, 1998, the research vessel *Aurora Australis* sailed from Hobart, Tasmania, in southern Australia to carry out a survey in the Antarctic sea ice zone. The aim of the survey was to study the Mertz Polynya in the middle of winter, when sea ice growth and bottom water formation are at a maximum. Taking accurate oceanographic measurements of the water temperature and salinity under sea ice, in an active polynya close to the coast of Antarctica, is something that can only be

---

[21] Broecker (1991).

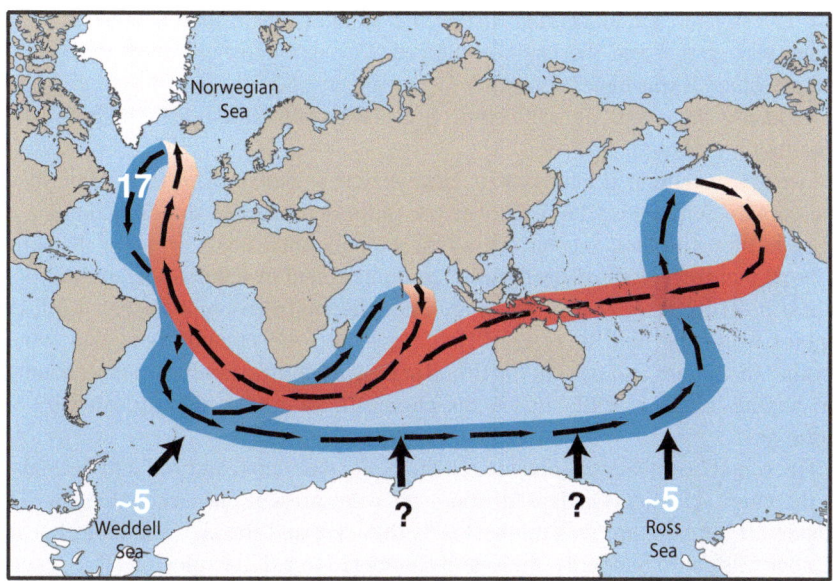

Fig. 5.3 The "great ocean conveyor" based on Broecker (1991). Bottom water is formed in the polar seas; it sinks to the ocean floor and flows into the Indian and North Pacific Oceans before returning to complete the loop in the North Atlantic. Numbers indicate estimated volumes of bottom water production (Johnson 2008) in "Sverdrups" (1 Sverdrup = 1 million cubic meters per second). Volumes of North Atlantic Deep Water (NADW) and Antarctic bottom water (AABW) are expected to be equal at around 17 Sverdrups although the exact sources of AABW are unknown

done from a ship. This work was part of a major scientific program of Australia's Antarctic Cooperative Research Centre (CRC), based in the University of Tasmania.

For those who are bothered by seasickness, a voyage that takes you into the sea ice zone is a blessing. This is because sea ice effectively decouples the wind from the ocean surface. Since surface waves are caused by wind, the sea ice zone is blissfully calm, no matter how strong the wind blows. There may be a raging storm creating huge waves out on the open ocean, but in the sea ice zone, the seas are calm. It is a great relief to leave the open ocean behind you and sail into the sea ice zone (provided, of course, that you are on board an icebreaker!).

The RV *Aurora Australis* is an icebreaker operated by the Australian government's Antarctic Division. Like all icebreakers, the ship's bow is designed to ride up over the floating ice. It is the weight of the ship riding over the ice that cracks it open and pushes it aside, allowing forward progress (not the sharp prow of the ship, as one might think). Powerful engines are needed to literally drive the ship over the ice, lifting the bow out of the water and onto the ice, cracking the ice apart and pushing forward.

The ship arrived at the edge of the sea ice zone on July 21, and the powerful main engines were changed to icebreaking mode. At 02:30 on the morning of July 22, *Aurora Australis* was about 180 km off the coast of Antarctic when an alarm sounded alerting the bridge that something was amiss in the engine room. The engineer on duty investigated the alarm and discovered a small fire in the engine room. The bridge was alerted and the engines stopped. Fire extinguishers were used, and at first it seemed it was under control.

Suddenly there was an explosion. A fireball swept through the engine room engulfing it in flames.

Fire alarms sounded and *Aurora Australis* crew were woken from their sleep. The captain sent a Mayday call over the radio and ordered the 54 scientists on board to report to their muster station on the helicopter deck. Some of the ships, 25 crew members donned firefighting equipment and evacuated the engine room to seal it off from the rest of the ship. Once the engine room was sealed off, the captain would be able to release Halon gas into the compartment that would smother the flames and put out the fire. But first all of the crew had to be accounted for; anyone left accidently inside the engine room would be suffocated by the Halon gas.

The scientists mustered, outside, on the helicopter deck, and roll call was taken by the chief scientist, Dr. Ian Allison – all were present and accounted for. The outside air temperature was minus 13 °C (8.6 °F), and the sky was dark, but the weather was calm. Suddenly, the vessel shuddered to a dull explosion. The generators and electrical systems shut down. The lights went out, and an eerie silence fell over the ship.

The *Aurora Australis* came to a dead stop at latitude 63° south, near the Antarctic Circle, in total darkness, powerless and drifting. With all the crew accounted for, at 02:52 the captain released the Halon gas to extinguish the fire. Would the fire be extinguished? Or would the order to abandon ship be given? The minutes ticked by in total silence.

By 03:00 the engineers had started an emergency generator and power was restored. The lights came back on, and the crew prepared the lifeboats and swung them outboard, ready to be lowered. The sea ice was thin enough that lifeboats could be used. Thick sea ice would render the lifeboats useless to escape from a burning ship since they could not be moved away and their fiberglass hulls would simply melt in the heat.

Temperature in the engine room began to fall. The Halon gas had worked and the fire was out. *Aurora Australis* would survive to sail again.

Later that day, once the engine room had cooled down and the poisonous gasses had been pumped out, it was possible to inspect the damage. Dr. Allison described the scene: "plastic insulation was melted and wires fused, control boxes with electronics were totally blackened and twisted, all lights were shattered and fluorescent light fittings hung in molten pendants from the ceiling."[22] One of the engines was able to be repaired by the ship's engineers, and after 3 days of hard work, *Aurora Australis* was able to limp back to Hobart under her own power. Nobody was injured, and it is enough to say that the incident has now become a historical footnote in the annals of Antarctic marine research.

---

[22] Allison (1999)

Luckily, oceanographic research is usually not as eventful as *Aurora Australis'* 1998 expedition to the Mertz Polynya. But the story serves to remind us that we must never take the ocean for granted. Collecting data in remote places like the Mertz Polynya is essential for us to measure and understand the workings of the great ocean conveyor.

All work at sea contains an element of risk that must never be forgotten. There are around 420 research ships with a length over 25 m operated by 49 maritime countries, including ships specializing in oceanography, fisheries, polar, and military research. According to statistics reported in the European Marine Casualty Information Platform, the average research vessel in 2016 was 60 m in length and 24 years old.

Back in Australia, an investigation into the cause of the fire found that it was due to a faulty fuel hose. The leak had spilled fuel oil that came into contact with the engine which then started a fire. It took several months to repair all the damage and get the *Aurora Australis* back at sea again. The ship's busy schedule was thrown into disarray, and several research voyages had to be cancelled.

One of them was mine!

***

In 1998, I was the leader of the Antarctic Cooperative Research Center's paleoenvironment program based in Hobart Tasmania. Our team carried out investigations into the history of the Antarctic ice sheet and the Antarctic marine environment. Antarctic research is an international endeavor, and our team had joined up with a group from Italy's national geophysics research institute, Istituto Nazionale di Oceanografia e di Geofisica Sperimentale (OGS), for a joint Antarctic expedition. Our original plan was to use the *Aurora Australis* in the summer of 1998–1999, but because of the fire, our expedition was delayed, and we secured the services of the New Zealand Research Vessel *Tangaroa* for the summer of 1999–2000 instead. Ours was to be the first Antarctic expedition for the *Tangaroa*, and it was also the first joint Australian-Italian Antarctic expedition ever undertaken.

A missing piece of information for understanding the Great Ocean Conveyor are the processes governing the formation of Antarctic bottom water. Although it is known that sea ice factories play a key role, their location around Antarctica and the quantities of bottom water they produce are a mystery. In the late 1990s, the famous American oceanographer, Wally Broecker, suggested that Antarctic bottom water production may not be in a "steady state," possibly because the sea ice factories are not stable.[23] To not be in a steady state would mean that the formation of bottom water along the margin had to switch on and off episodically, from place to place. But why would that happen? And how could we hope to find an answer?

We planned our expedition of the *Tangaroa* to the visit the George Vth Land coast, where a polynya was formed on the lee side of the floating tongue of the

---

[23] Broecker et al. (1998).

Fig. 5.4 (**a**) Sea conditions in the Mertz polynya viewed from the bridge of RV Tangaroa with 70 knot winds. (**b**) The author at Mawson station, Antarctica in 1997. The tiny red ship in the background is the Aurora Australis. (**c**) Sediment core of siliceous mud and diatom ooze (SMO) collected from the seafloor of the Mertz polynya. (**d**) The floating terminus of the Mertz Glacier; (**e**). The author pictured with Dr Gene Domack (center) and Dr Phil O'Brien (at right) on the beach of Heard Island. All photos by the author

Mertz Glacier. The George Vth Land coast was made famous by Douglas Mawson's 1911–1913 expedition, where two of his companions, Dr. Xavier Mertz and Lt. B. E. S. Ninnis, perished. Two of the largest outlet glaciers in the region bear their names.

Oceanographers had recently discovered that the Mertz Polynya was an important source of bottom water, generating about one Sverdrup of Antarctic Bottom Water; in other words about one-seventeenth of all Antarctic Bottom Water is produced in this one polynya. There are two main factors that combine to make the Mertz Polynya an important regional sea ice factory and location for bottom water formation. First is the solid wall of ice, rising 300 m (1000 feet) from the ocean surface, made by the floating terminus of the Mertz Glacier (Fig. 5.4). The Mertz extends over 60 km (37 miles) offshore from the coast and across the shelf to where it grounds on a shallow bank on the outer shelf. The westward-flowing prevailing current along the coast piles up ice on the eastern side of the Mertz Glacier leaving an open area on the leeward, western side (where the polynya is).

The second factor is the strong offshore winds that blow from this coast. These winds blow steadily and at phenomenal speeds as first noted by Douglas Mawson at his camp in Commonwealth Bay. The reason is because cold, dense air masses continuously descend from the Antarctic ice sheet down the valley of the Mertz Glacier and blast their way out to sea ("katabatic" winds). On board *Tangaroa*, we lost several days working due to these winds that exceeded 70 knots (130 km/hour, over hurricane force) while we were there; 70 knots is hurricane force wind and is as high as the wind gauge on *Tangaroa's* bridge instruments was able to read!

For our expedition, we came equipped to map, sample, and photograph the continental shelf. A previous expedition to the Mertz Polynya region had discovered a large sediment drift deposit located in a shelf valley about 850 m (2800 feet) deep, and we hoped that careful study of it would tell us something about the history of bottom water formation. We arrived in the Mertz Polynya in February 2000 and began our work.

Samples of Antarctic shelf sediment that we collected are composed of siliceous mud and diatom ooze, called "SMO" for short. SMO has a dark greenish color with black lamina, and it bears an unfortunate resemblance to mucus. Sediment samples are collected using a piston corer that is lowered to the seabed and which penetrates into the sediment. Sediment core samples are collected in lengths of plastic tube inserted into the steel pipe of the piston corer. Successful cores yield a plastic tube filled with SMO.

Once back aboard ship, the plastic, SMO-filled tubes were carefully taken below to *Tangaroa's* laboratory where they were split lengthwise using a special core-cutting device. The core cutter consists of two saws, with opposing blades that cut only the plastic tube without disturbing the sediment inside. The plastic core tube is pushed along lengthwise leaving two cuts, one on each side. Then a thin wire is drawn along the length of the core to slice the sediment neatly into two, long, half-sections.

The opening of a core sample is one of the most exciting moments in the life of a marine geologist. The first sample collected on any scientific voyage often attracts a good number of curious crew members who were off-watch. We all gathered around to witness the first "grand opening" of our first sediment core. The two halves of the core were prized apart to reveal the length of greenish, slimy gooey sediment which gave off an unpleasant odor of rotten eggs. The crew members rapidly withdrew in disgust, mystified by the excited comments and exclamations made by the scientists who were closely scrutinizing what appeared to be a tube of reeking green muck!

Twenty-five core samples were collected during the expedition, and as they were opened, one by one, the scientific party were on occasion overcome with enthusiasm, and there could be heard singing in the lab: "oh there's no business like SMO business there's no business I know"! The crew members remained unimpressed by either our sediment samples or singing ability. They did however appreciate the fantastic coffee that was brewed by our Italian colleagues every day in our lab, the aroma of which more than compensated for the occasional whiff from our core samples.

The mariners who crew on board research vessels are drawn from the merchant navies of their countries. They learn on-the-job about ocean research, and on the (mostly Australian) ships I have worked, the crew refer to their scientific passengers as "boffins." I generally accept this as a term of endearment even if it is not always intended that way. I have met crew members who wrote poetry and music, some are artists who paint pictures or carve figures out of wood, all during their time off-watch. I have met sailors who were once navy test pilots sailors who have spent their whole lives at sea and others who came to work on ships later in their lives. Nearly all are philosophers, which must go hand-in-hand with a life at sea. I have a deep respect for those professional "scientific" sailors, who make research at sea possible and who have helped all of us boffins to safely complete our work.

Although it is sometimes possible for core samples to reveal key information just from what can be seen upon opening, it is more often the case that detailed laboratory analyses of the sediments are needed to measure their properties, age, and organization (fossil content, thickness of layers, evidence of current bedding, etc.). Data collected from different laboratory experiments are necessary before an interpretation can be made. This was the case for our sediment cores, and it was many months later that the story became clear.

Strong bottom currents had indeed been switching on and off over the last several thousands of years, as recorded by cross-bedded layers of sediment containing greater quantities of sand (strong currents) and laminae of fine mud and diatoms (weaker currents). We hypothesized that over decades and centuries, the Mertz Glacier calved off a large iceberg, and the Mertz Polynya shrank in size (or disappeared) until conditions favorable to its restoration reoccur.[24] This explains why the production of Antarctic Bottom Water is not in a steady state, but rather it fluctuates over the centuries, sometimes greater and sometimes lesser in volume.

Proof that our interpretation was correct came on February 12–13, 2010, when the Mertz Glacier calved off and the Polynya was dramatically reduced in area, losing a portion of its capacity for producing bottom water. Measurements and modelling showed that bottom water was reduced by up to 23%.[25]

This example highlights how highly sensitive Antarctica is to environmental changes and how the glacial-oceanic systems are interlinked. The calving of glaciers and floating ice shelves reported in the news has implications for the production of bottom water that drives the global conveyor. This is not just about the transport of heat around the globe. The conveyor also transports one vital element that is essential to life in the ocean: oxygen.

---

[24] Harris et al. (2001).

[25] Kusahara et al. (2011).

Chapter 6
The Lungs of the Ocean

Mother, mother ocean, I have heard you call
Wanted to sail upon your waters since I was three feet tall
You've seen it all. You've seen it all.
Jimmy Buffet
A Pirate Looks at Forty, Dunhill Records, 1974

Abstract The polar seas are the lungs of the ocean. When sea ice forms, it makes cold and salty water that is enriched in oxygen. Polar seas are where the oceans "breathe in." Oxygen dead zones are created where rivers polluted by fertilizer reach the ocean and are a growing concern globally. Carbon dioxide is also absorbed by the ocean along with oxygen in the polar seas. In seawater carbon dioxide makes weak carbonic acid causing ocean acidification, the "evil twin" of global warming. But the oceans have not always had lungs that work the way they do today. In the geologic past, the deep ocean waters were hot, salty, and sometimes deprived of oxygen. We will also learn about a time 6 million years ago when the Strait of Gibraltar was closed off and the Mediterranean Sea nearly dried up.

Keywords Dissolved oxygen · Antarctic Bottom Water · North Atlantic Deep Water · Equatorial upwelling · Anoxic · Photic zone · Oxygen minimum zone · Acidification · Messinian Salinity Crisis

Plants are the source of all the oxygen in our atmosphere, produced by photosynthesis. Plants on land produce about half of the Earth's oxygen supply, and phytoplankton in the oceans produce the other half. Oxygen is consumed by all animals that breathe, by the decomposition and combustion of organic matter, by the oxidation of minerals, and by burning fossil fuels. Over geologic (deep) time, the amount of oxygen produced is balanced by the amount consumed so that the concentration in the atmosphere remains constant at around 21%. But the ocean-atmosphere system is dynamic; it is continuously changing with the seasons and with regional patterns

© Springer Nature Switzerland AG 2020   81
P. T. Harris, *Mysterious Ocean*, https://doi.org/10.1007/978-3-030-15632-9_6

of primary production and absorption and consumption of oxygen. It is such a complex system that the exact numbers involved in the global oxygen budget are not exactly known, and probably never will be.

The greatest fluxes of oxygen into and out of the oceans take place mainly at high latitudes, above 60° north and south. Outgassing (release of oxygen from the ocean to the atmosphere) is related to seasonal blooms of phytoplankton that occur during spring and summer. The oceans' contribution of oxygen to the atmosphere oscillates back and forth, from the northern to the southern hemispheres, with the changing seasons.

Ingassing (adsorption of oxygen) occurs where surface waters absorb oxygen from the atmosphere. Oxygen dissolves in cold seawater better than warm seawater. In the cold, polar oceans, dissolved oxygen reaches concentrations of as much as 9 mm/l; warm seawater in the tropics (at temperatures above 25 °C) can only absorb around 4 mm/l. One key difference between the poles and tropics is that polar oceans "breathe in" in winter; at the same time, bottom water is being formed. At the same time that the cold polar waters are recharged with oxygen, they sink below the surface to be replaced by warmer surface water. Oxygen-charged bottom waters, produced by the polar ocean "lungs," flow into the deep ocean and around the ocean basins following the great ocean conveyor, bringing oxygen to animals living in the depths of the abyssal ocean.

Oxygen is slowly depleted in bottom water traveling along its path across the abyssal ocean floor (Fig. 6.1), as it is consumed by fish, benthic animals, and bacteria. North Atlantic Deep Water flowing southward through the Atlantic arrives in the Southern Ocean with less than half of its original concentration of dissolved oxygen. Even greater is the depletion of dissolved oxygen from Antarctic Bottom Water as it makes its slow journey into the North Pacific. By the time bottom water reaches the deep basins near Alaska, it contains less than 1 mm/l of dissolved oxygen (it is nearly completely depleted of oxygen).

The conveyor loop is closed where cold, oxygen-depleted, bottom waters are brought to the surface, a process that oceanographers call "upwelling." The polar seas are locations where the ocean "inhales" oxygen, and upwelling areas are where the ocean "exhales." Upwelling zones are among the most productive parts of the oceans. The cold upwelled water is depleted in oxygen but is enriched in carbon dioxide and loaded with nutrients, phytoplankton food. It is not a coincidence that many of the world's best fishing grounds are associated with upwelling zones.

But what causes upwelling to occur?

Upwelling happens in different places of the ocean mainly for two different reasons: diverging currents and offshore winds. The most important way to make an ocean upwelling is by winds. In some situations, offshore wind blows the water away from the land, and this draws cold water up from the depths to replace the water that is moved offshore. But the most important upwelling process involves wind blowing along the coast (parallel to the coastline). The wind transports the water at right angles to the wind direction by Ekman transport as we learned in Chap. 5. This can either cause water to pile up on the coast or to push it away from the coast. In the latter case, this causes upwelling.

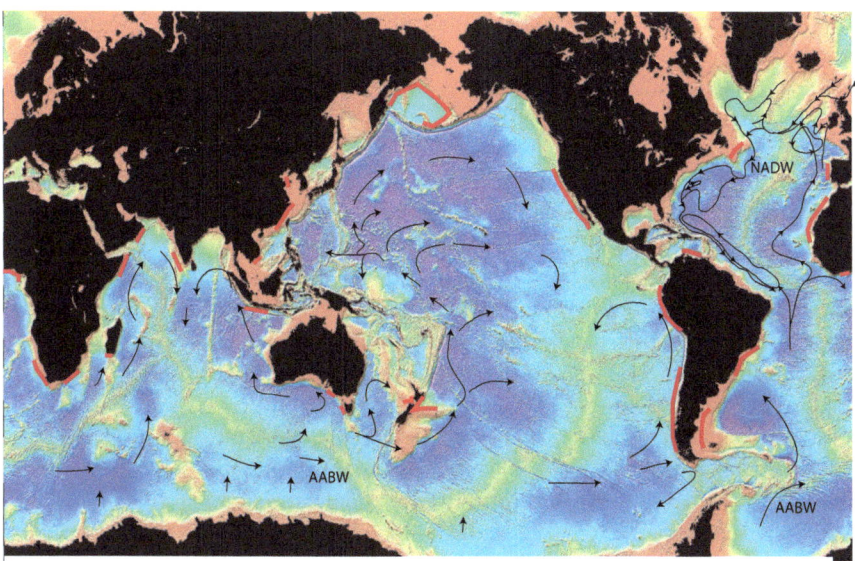

Fig. 6.1 Bathymetry of the world ocean (from Smith and Sandwell 1997) with arrows showing global bottom water circulation in the ocean. Deep circulation is driven by southward flow of North Atlantic Deep Water (NADW) formed in the Norwegian and Labrador Seas and the northward flow of Antarctic Bottom Water (AABW) formed in the Ross Sea, Weddell Sea, and along the East Antarctic continental shelf. NADW flow arrows are based on Schmitz and McCartney (1993) and AABW water flow arrows are based on Tchernia (1980), reinterpreted based on bathymetry of Smith and Sandwell (1997). Red lines show major coastal upwelling zones (from Kämpf and Chapman 2016). Not shown is equatorial upwelling

Divergent ocean currents also cause upwelling to occur. Remember how we found that warm moist rising air at the equator gets deflected by the Coriolis force so that the wind actually blows toward the east, along the equator (the "easterlies")? Wind shear on the surface of the ocean sets the ocean in motion, and the moving water is also deflected by the Coriolis force, to the north on the northern side of the equator and to the south on the southern side of the equator. The result is an ocean current divergence along the equator. Surface water moving away from both sides of the equatorial divergence draws cold water up from the depths, and voila, the result is equatorial upwelling!

Upwelling along the equator produces a band of nutrient-rich water that is irresistible to phytoplankton. Extending across the equatorial Pacific from South America to the island of Papua New Guinea and across the South Atlantic from the coast of equatorial Africa to Brazil, equatorial upwelling can be seen from space as discrete bands of high surface chlorophyll concentration.

Oxygen is carried into the deep oceans by bottom waters to support all the animals that exist in the deep sea. In some locations, the available oxygen is completely used up resulting in a condition known as hypoxia. In severe cases, hypoxic waters can become an ocean "dead zone" since no life that depends on oxygen can exist in such environments. This situation can happen where rivers discharge into the ocean carrying large amounts of nutrients from the land which are then consumed by marine plants, animals, and bacteria.

Tiny, single-celled plants float in the surface waters where they consume nutrients and produce oxygen via photosynthesis. When they die, the dead plants sink to the seabed where they are consumed by benthic animals and bacteria, which consume oxygen. In this way, the bottom waters can become oxygen starved (hypoxic or anoxic[1]) even while the surface waters contain abundant oxygen.

Oxygen is continuously exchanged between the atmosphere and surface waters. Waters in the photic zone (where sunlight penetrates) can produce excess oxygen from photosynthesizing planktonic algae. In other places where planktonic algae are not abundant, surface waters may absorb oxygen from the atmosphere to achieve equilibrium.

Below the photic zone (i.e., below around 200 m depth), the only source of oxygen is mixing downward from the surface by density currents (cold, salty water sinking) or by stirring of the water by the wind, tidal currents, or storms. Below this depth, physical mixing processes are unable to supply oxygen faster than it is consumed by animals and oxidation of organic matter. The result is the formation of an oxygen minimum zone, sandwiched between the surface ocean and the deep ocean. It is normal for a vertical profile of oxygen to be high in the upper 200 m of surface waters (the mixed layer) with a zone depleted in oxygen from 200 to 300 m depth then gradually rising again with increasing depth.

In some locations, the depletion of oxygen is more dramatic. In the depths of the Black Sea and the Baltic Sea, for example, the basins are cut off from the ocean by shallow sills, and there is no supply of oxygen into the deeper water. The entire water column below the surface is consequently starved of oxygen. Storms in these sulfide seas can bring hypoxic water into shallow depths killing the fauna and releasing hydrogen sulfide gas (the "rotten egg" smell).

Humans have increased the number and extent of ocean dead zones in recent decades because of growing populations and the use of fertilizers in agriculture. Prior to the industrial revolution, the two main ingredients in fertilizer, phosphate and nitrogen, were hard to come by. Growing food was limited by the cost and availability of fertilizer. Bird manure (guano) was mined as a source of phosphate and nitrogen fixation occurred only slowly, by natural processes.

Nitrogen fertilizer became much easier to produce with the invention of the Haber process in 1908. This industrial catalysis process requires high pressures and temperatures and uses natural gas as a hydrogen source and air as a nitrogen source. Suddenly, fertilizer became plentiful and cheap. Widespread and excessive use of fertilizer to grow crops is the root cause of expanding ocean dead zones. The excess nutrients make their way into rivers that eventually reach the ocean to be used by marine life, which also use up the available oxygen. In this way, humans have

---

[1] "Hypoxic" means depleted of oxygen; "anoxic" means total absence of oxygen. Anoxic sediments typically have the smell of rotten eggs because the bacteria use sulfide to process organic matter instead of oxygen.

altered the balance between the rates of production and consumption of dissolved oxygen in the ocean. Dead zones now appear regularly in the Gulf of Mexico (at the mouth of the Mississippi River) in Chesapeake Bay, in the Kattegat Strait (at the mouth of the Baltic Sea, between Denmark and Sweden), and in the Adriatic Sea.

***

Oxygen is not the only gas in the atmosphere that mixes into the ocean affecting marine life; another one is carbon dioxide (CO_2). When CO_2 dissolves in seawater, the net effect is to decrease the carbonate (CO_3^{2-}) and increase the concentration of hydrogen ions (H+) thereby lowering pH, making the oceans more acidic. The more carbon dioxide in the atmosphere, the more it mixes into the ocean lowering the pH. Carbon dioxide dissolves more easily into cold water, like oxygen, and it follows a similar pathway through the deep ocean, inhaled at the poles and exhaled in upwelling zones.

Present-day carbon dioxide levels are over 400 parts per million (ppm), which is ~120 ppm greater than the preindustrial value of ~280 ppm. The oceans have absorbed around 30% of the carbon dioxide humans have emitted (the atmosphere would already contain over 500 ppm of CO_2 if it weren't for the ocean).[2] The increase in ocean CO_2 has caused the average ocean surface pH to drop by 0.1 pH unit, from pH 8.25 to 8.14. Although this change may seem small, it amounts to an increase of 30% in acidity because the pH scale is logarithmic (pH 7 is ten times more acidic than pH 8).

The acidification of the oceans has dire consequences for marine life because it increases the energy needed to build shells out of calcium carbonate. Calcium carbonate ($CaCO_3$) is dissolved in seawater in abundance, but many organisms, such as calcareous plankton, bottom-dwelling mollusks, echinoderms, and corals, find it increasingly difficult to build their shells with lowering pH. For this reason, ocean acidification is sometimes called the "evil twin of global warming."

In the upper 200 m mixed layer, dissolved gas content is in equilibrium with the atmosphere. In other words there is no excess capacity to store CO_2 in the upper 200 m because that volume is already taken with CO_2 as it mixes down into the surface ocean. However the volume of ocean below the mixed layer has been out of contact with the atmosphere for some years. This deep water is not in equilibrium with the current atmosphere and does have spare capacity to absorb more CO_2.

In places where bottom water is formed, CO_2 dissolves into the water, and by sinking into the deep ocean, it is stored out of contact with the atmosphere. This is the main way that the ocean is now absorbing excess CO_2 and sequestering it into the deep ocean. The world faces two dilemmas, one far in the future when the bottom water resurfaces and another if the sea ice factories are closed down by human-caused climate change.

I mentioned earlier that the age of deep ocean water is about 1000 years on average. In other words, water at the bottom of the ocean was last in contact with the atmosphere 1000 years ago, when the atmospheric CO_2 concentration was around

---

[2] Doney et al. (2009).

280 pp. If in the next few centuries we manage to reduce our emissions and gradually reduce atmospheric CO_2 to level around say 300 ppm, those future generations will be cursing us because upwelling bottom water will be releasing excess CO_2, stored under current levels of 400+ ppm CO_2, back into the atmosphere when it is upwelled from the deep ocean. Future generations will have to cope with higher CO_2 even with zero human emissions in order to account for the excess CO_2 we are pumping into the deep ocean today.

The second factor is the possibility of a slowdown of the polar sea ice factories. As we warm the Earth and oceans, the area of sea ice production is shrinking. Melt water from Greenland and Antarctica may slow or stop the Arctic bottom water factory because if the surface water is too fresh to start with, the formation of sea ice does not make it salty enough to sink below the existing water mass. The combination of these factors means that the rate of bottom water formation is slowing. Experts predict a reduction in bottom water production in the next decades and slowing of the global ocean conveyor[3] which translates into a reduction in the oceans' capacity to absorb CO_2 and sequester it into the deep ocean. The rate of CO_2 addition to the atmosphere may then increase even if humans reduce greenhouse gas emissions.

***

Oxygen in the present-day oceans is sourced from polar seas where sea ice is formed every winter, and these areas are the "lungs" of the ocean. These are the sources of oxygen-rich bottom waters that supply the abyssal depths for the global ocean.

But this has not always been the case.

During the Cretaceous period, when dinosaurs roamed the world, the climate was much warmer than today, and there were no polar ice caps and no seasonal sea ice factories existed. And we know from sediment deposits that the Tethys, Early Atlantic, and eastern Pacific Oceans were at least partially anoxic for brief (about 1 million years' duration) periods of time, around 120 million years ago and again around 97 million years ago.

How could an entire ocean become starved of oxygen? Could it happen again?

Although we have no anoxic oceans today, we do have anoxic seas like the Baltic and Black Seas. So, we can make some educated guesses at the conditions needed to make an ocean turn anoxic. Clearly the Earth must have been a lot warmer, probably with higher carbon dioxide levels than today. If sea ice factories did not produce bottom water, what process did drive deep ocean circulation? One plausible theory involves bottom water production driven by evaporation to produce warm, high-salinity bottom water.[4]

Seawater that is coldest and saltiest is most dense. But what if the coldest waters at the poles were not that cold but the saltiest waters were a lot saltier. If polar waters never got colder than 9 °C, for example, then seawater that has 35 ppt salt has a

[3] Jackson et al. (2016).

[4] Friedrich et al. (2008).

density of 1.027 g/ml. When warm water gets very salty, it also becomes very dense. Seawater at 25 °C and a salinity of 40 ppt also has a density of 1.027 g/ml. What if, in this very hot Earth scenario, there are shallow, semi-enclosed seas next to deserts with no rivers flowing into them and high evaporation rates? Could these shallow seas become a kind of factory for hot salty water?

This is the exact situation in Shark Bay, Australia, at the present time. Seawater salinity reaches an astonishingly high 70 ppt, due to the hot climate and desert conditions. This salinity is so great that even at a temperature of 30 °C, its density of 1.048 g/ml is enough to sink *below* Antarctic Bottom Water. Shark Bay does not produce enough warm, hyper-salty water to replace Antarctic Bottom Water in the oceans today. But the situation was different in the Cretaceous because there were a lot of areas like Shark Bay, pumping out dense, warm salty water. And because there were no other sources of bottom water, the entire global circulation system moved at a much slower pace.

Since there were no polar ice caps in the Cretaceous, global sea level was more than 100 m (300 feet) higher than today, and the oceans flooded areas of low-lying continents, making vast, shallow, continental seaways. At this time, the center of Australia was flooded, and there was a *Cretaceous Seaway* extending over the middle of what is now the central United States and into Canada (think of flooding the entire Mississippi River system under a shallow sea!). High evaporation rates produced vast quantities of warm, extra-salty water that was dense enough to fill the ocean basins. Warm water is unable to store as much dissolved oxygen as cold water, and because the ocean conveyor turned much more slowly, all the oxygen was consumed, and some ocean basins became anoxic.

There were probably other factors involved in turning the oceans anoxic. Primary production must have been very effective at producing abundant organic matter that consumed oxygen during its decomposition once it settled to the seabed. The continents had to be arranged in such a way that the sources of warm, salty bottom water dominated in certain ocean basins. The Tethys Ocean, for example, straddled the equator and became encircled by land. The early North Atlantic also had limited exchange with the world ocean, being cut off by encircling land areas. Subtle changes in continental configurations, combined with climate and biological systems, allowed anoxic conditions to persist for a brief (in geologic terms) few millions of years.

And this leads us to an amazing revelation; life in the oceans at any given time is utterly dependent upon the particular, random configuration of the continents. For most of geologic time, the continental configuration has allowed for ocean circulation to ventilate the deep ocean with oxygenated water. But on some rare occasions, the configuration of the continents has combined with the climatic conditions and prevailing dominant biota to give rise to anoxic ocean basins which most benthic animals cannot tolerate.

Could anoxic oceans occur again in the future? Could anthropogenic climate change cause the oceans to become anoxic with the present configuration of continents? The answer is, yes, it is possible for at least parts of the ocean. First of all, it must be recognized that about 5% of the ocean volume, especially in the

North Pacific, is already depleted of oxygen (below 1.6 mm/l). If we do not stop adding more and more carbon dioxide into the atmosphere, warming of the oceans will reduce sea ice production, and hence the lungs of the ocean could falter. The result would be less oxygen supply to the deep sea which could cause parts of the oceans to turn anoxic.

One of the most dramatic events in the relatively recent geologic past was the so-called Messinian Salinity Crisis in the Mediterranean Sea. Remember how the Tethys Ocean was squashed out of existence by the colliding continents that split away from Gondwana? The Mediterranean Sea is the last remnant of the once mighty Tethys Ocean, and its only connection with the Atlantic and the rest of the world ocean is through the narrow Strait of Gibraltar. Around 6 million years ago, tectonic forces pushed Africa against Europe, and for a brief time, the Strait of Gibraltar was closed.

The rivers that flow into the Mediterranean do not provide enough water to balance the amount that is evaporated. The result was that, after a few thousands of years, the Mediterranean nearly dried up! When the Strait of Gibraltar reopened, the Atlantic flooded back in and refilled the Mediterranean. Then the Strait closed and the Mediterranean was cut off again. This happened over and over, for a period lasting around 7000 years, known as the Messinian age between about 6 and 5.3 million years ago.

Each time the basin dried up, thick layers of salt, gypsum, and other minerals were deposited on the seafloor. About 1 mm of salt is deposited for very meter of seawater evaporated, and beds of salt many meters in thickness were deposited each time the water was replenished. At the peak of the "crisis," the Mediterranean may have completely dried up, leaving a valley floor up to 5000 m below sea level interspersed with a few, Dead Sea-like lakes. Calculations suggest that adding the volume of the Mediterranean Sea to the rest of the world ocean would raise global sea level by around 10 m. The flow of water through the Strait of Gibraltar, each time it reopened, would have been spectacular to see, if only humans had evolved in time to witness it!

The arrangement of the continents can give us warm salty bottom water and anoxic oceans as well as desiccated ocean floors when they become exposed. But that's not the only thing that is controlled by the configuration of the continents. The configuration of the continents affects the Earth's albedo, the production of bottom water, and the flow of ocean currents. It even determines if the Earth is poised to enter an ice age.

It seems that there are three continental configurations that favor the world entering an ice age: (1) lack of continents that deflect warm ocean currents into polar regions such as might occur if most continents (or a supercontinent) were located along the equator; (2) a continent located at the pole like Antarctica; and (3) an ocean at the pole that is nearly landlocked, like the Arctic Ocean. So, as you can see, the Earth is presently scoring two out of three of these factors and has been poised in ice age mode for the past 40 million years.

What actually causes an ice age to begin and what causes it to end? The last ice age ended 10,000 years ago, so are we headed for another one soon? We shall look at the answers to these questions next.

Chapter 7
Frozen Ocean: Ice Ages and Climate Change

"As the axial tilt increases, the seasonal contrast increases so that winters are colder and summers are warmer."
Milutin Milanković
Canon of Insolation of the Earth and Its Application to the Problem of the Ice Ages, 1941

Abstract The world has two major ice sheets – in Antarctica and Greenland – but their histories are completely different. The Antarctic ice sheet evolved over 30 million years ago when South America separated from Antarctica to create the Drake Passage. This allowed the Circumpolar Current to form, isolating Antarctica and turning it into the coldest, highest, and driest continent. Greenland is the last major remnant of continental ice sheets that have grown over large parts of North America and Europe repeatedly for the last 2 million years. In this chapter, we will meet Milutin Milanković and learn about his theory for ice ages. Ice sheets over Europe and North America lowered sea level by 130 m, and when they melted, there were huge floods in Washington State in the west, as well as in Eastern Canada. The rising sea level had many consequences: it refilled the Black Sea, possibly explaining the biblical flood story. Rising sea level flooded the Gulf of Carpentaria in Australia and the Persian Gulf Oasis. All the fish, kelp, and corals living on the continental shelves today, including the Great Barrier Reef, are recent arrivals that only moved in over the last 10,000 years or so.

Keywords Antarctic glaciation · Ice-rafted debris · Milutin Milanković · Glacial erratic · Pleistocene · Ice age · Milanković theory · Solar insolation · Ice sheet · Ice shelves · Sea ice · Missoula Floods · Lake Agassiz · Younger Dryas · Black Sea · Noah's flood · Tidal range · Great Barrier Reef

The continent of Antarctica straddles the South Pole, and it has remained in that position for the last 170 million years. For most of that time, Antarctica's climate has been quite mild. In the Cretaceous, the continent was covered in forest including the deciduous angiosperm *Nothofagus*. Ferns grew, and dinosaurs dominated the food web. Dinosaurs became extinct when a large meteor hit the Earth around 65 million years ago, but the Antarctic climate remained temperate. Then, around 30 million years ago, Antarctica's climate began to cool down. Glaciers formed in the mountains, and gradually, they grew and extended down to sea level. Eventually, the entire continent was covered in an ice sheet 4 km (2 miles) thick.

Why Did Antarctica Become so Cold?

The fragmentation of Gondwanaland involved Africa, South America, India, and Australia all splitting off and moving north. Antarctica remained alone, fixed at the South Pole, but the climate was kept warm by the circulation of warm ocean currents. As the continents moved north, the ocean area around Antarctica began to expand. India and Africa had split off leaving only South America and Australia connected to Antarctica. At this stage, around 36 million years ago, the Brazil Current flowed south along the margin where Argentina (Tierra del Fuego) was then joined to the Antarctic Peninsula, and the East Australian Current flowed south along the east coast of Tasmania and into what is now the Ross Sea. These currents carried warm waters into the coastal areas of Antarctica, keeping the climate seasonally warm and temperate, just as the Gulf Stream keeps the climate of Norway temperate today.

But the land connections between Tasmania, Tierra del Fuego, and Antarctica were gradually stretched and finally broken as the continents moved apart. Tasmania broke away first, followed by South America. Drake Passage is the name given to the opening between the southern tip of South America (Tierra del Fuego) and the Antarctic Peninsula, and its opening spelled the end for Antarctica's temperate climate. The warm ocean currents from Brazil and East Australia no longer reached as far south to warm the Antarctic. Most importantly, however, was the continuous expanse of ocean that surrounded Antarctica.

The only example today of an ocean current that flows unhindered around the globe is the Antarctic Circumpolar Current. This current flows from west to east, encircling Antarctica, and has the effect of isolating the continent from any warm ocean currents. It is because of the Circumpolar Current that the Antarctic glacial ice cap has persisted there for the past 30 million years. The reason we know this is because the Circumpolar Current and the Antarctic glacial ice cap came into existence at about the same time. And the reason we know this is because of fossils contained in sediments deposited on the ocean floor.

Fossils of planktonic plants and animals that are known to have lived in either colder or warmer climates reveal the dramatic changes that occurred 30 million years ago in the Southern Ocean around Antarctica. Prior to around 35 million years ago, the fossils indicate warm ocean conditions with colder conditions after that time. Furthermore, deep-sea sediments dated to after around 25 million years ago contain pebbles and large rocks that were picked up by Antarctic glaciers and carried out to sea by floating ice bergs. These pebbles and rocks (known as "ice-rafted debris") are found randomly embedded in what is otherwise fine-grained mud comprised of fossil shells of diatoms and planktonic foraminifers. Ice-rafted debris is a clear indicator of glacial conditions on the adjacent land mass.

At the same time that warm currents stopped flowing to the Antarctic coast, another process was at work cooling the entire Earth's climate. This was the uplift of the Himalayas caused by the collision of India with the Asian continent. The newly uplifted mountains contained silicate rocks like calcium feldspar which interact with atmospheric carbon dioxide to yield clay minerals. This process is known as "weathering," and it has an important effect on global CO_2 levels. At times of active global mountain building, the increased rate of weathering acts to reduce the carbon dioxide content of the atmosphere. This is what happened during the Oligocene epoch around 30 million years ago, which saw a long-term, gradual decline in atmospheric carbon dioxide levels.

The Antarctic ice sheet presently covers an area of almost 14 million square kilometers (5.4 million square miles) covering all but 2% of the continent. The ice sheet is over 2 miles (4 km) thick, and it contains 26.5 million cubic kilometers (6.4 million cubic miles) of ice. The ice sheet rests on the land (it is not floating), and its dome-shaped upper surface rises to an elevation of up to 4000 m (13,000 feet) above sea level. If it were melted, the Antarctic ice sheet would raise global sea level by around 60 m (200 feet). By comparison, melting the Greenland ice sheet would raise global sea level by around 6 m (20 feet). In other words, Antarctica holds ten times as much ice as Greenland. It is important to remember that when floating ice (sea ice and floating ice bergs) melts, it has no effect on sea level. Only melting ice that is resting on land causes sea level to rise. I recommend experimenting with ice cubes and glass of a fine malt whiskey to prove this fact to yourself!

Global climate change has caused both the Antarctic and Greenland ice sheets to lose mass at increasing rates. Measurements taken between 2002 and 2006 show that the Greenland ice sheet is melting at a rate of about 239 cubic kilometres per year, which is a sixfold increase over the previous decade.[1] The Antarctic ice sheet lost around 86 cubic kilometres per year over the period 2002–2015.[2] Together, melting of Greenland and Antarctic ice sheets is causing global sea level to rise around 1 mm/year.

The glaciation of Antarctica over the last 30 million years has intensely eroded the continent. All of the soft sediments and sedimentary rocks on the continent have been stripped away by the ice, leaving only the hard, crystalline basement rocks behind. The glacial ice has carved valleys up to 2500 m deep into Antarctica's continental bedrock. Drilling of the thick sedimentary sequences deposited where Antarctic glaciers debouched into the sea provides further evidence for the long history of ice sheets covering the continent.[3] The land-derived sediments surrounding the continent cover fully one-third of the Southern Ocean seafloor (Fig. 7.1), in a sediment drape that has an average thickness of nearly 900 m.[4]

---

[1] van den Broeke et al. (2017).

[2] Forsberg et al. (2017).

[3] Cooper and O'Brien (2004).

[4] Harris et al. (2014).

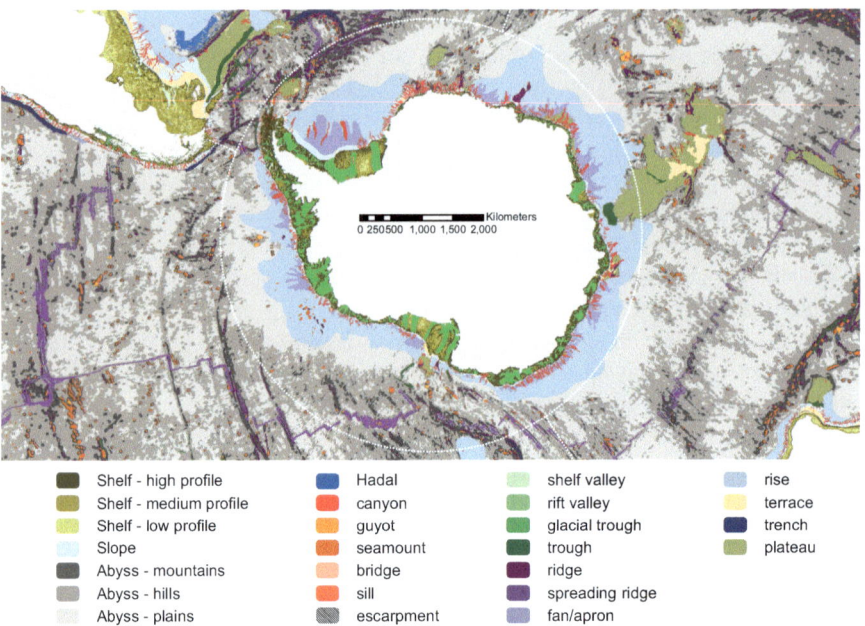

Fig. 7.1 Map of the ocean floor around Antarctica (Harris et al. 2014). Note the halo of sediment shown in blue (continental rise) which has been deposited around the continent, sourced mainly from glacial erosion of the Antarctic continent

The opening of the Southern Ocean and establishment of the Circumpolar Current are critical defining factors for the modern oceans. Geologically speaking, the Earth has entered a glacial phase that could easily slip into a snowball Earth scenario. This is because the formation of a continental ice sheet covering the whole of Antarctica affects the entire planet. It has a general cooling effect, and it created the conditions necessary for the next event in the ocean's history: the Pleistocene glaciations.

***

Twenty thousand years ago, the Earth was in the grip of an ice age that covered Northern Europe and most of North America under an ice sheet that was over 3 km (a mile and a half) thick. The ice sheets formed over a period of more than 80,000 years, expanding when winter snow fell and was not melted in the next summer, shrinking at times of warmer summers. As the ice sheets grew, the net result was the transfer of water from the oceans onto the continents causing sea level to fall. At the peak of the last ice age, 20,000 years ago, the mass of ice was so vast that global sea level was lowered by around 130 m (400 feet). The sea level was lowered enough to expose vast areas of continental shelves around most of the continents.

Louis Agassiz (1807–1874), a Swiss-born glacial geologist and later Harvard Professor of zoology and geology, proposed in 1837 that the world had once been in the grip of an "ice age," based on evidence of striations and grooves that he inferred were carved by moving ice. Based upon maps of these features plus glacial erratics (rocks dropped by glaciers), he concluded that ice had once covered Canada, the Northern United States, and Europe.

Evidence piled up hinting at the existence of widespread ice sheets at some time in the past. Information on glacial deposits (moraines) and erosion features (kettles, lakes, fjords, and other features) was collected throughout the nineteenth and early twentieth centuries. By working out the stratigraphy of moraines overprinting older glacial features, scientists proved that there had been at least four ice ages in the last 2 million years (the Pleistocene). But the reason why the ice ages had occurred was unclear; what had caused the ice sheets to grow and cover all of Northern Europe and North America, including all of Canada, Alaska, and the Northern United States? How long had the ice ages lasted? And then what had caused the ice sheets to melt away and for the ice age to end? Nobody knew the answers to these questions.

The cause of the Pleistocene glaciations remained a mystery until the 1970s, and the person who made the biggest contribution to solving the mystery did not live to see it solved. His name was Milutin Milanković (pronounced Mi-lan-ko-vitch), a Serbian engineer, mathematician, and astronomer. Milanković was born in 1879 in Croatia, and he studied engineering at the University of Vienna completing his PhD in 1904 on the subject of reinforced concrete building technology. By coincidence, Alfred Wegener also completed his PhD (in astronomy) that same year (1904), and the two became correspondents later in life.

Milanković was appointed Professor of Mathematics at the University of Belgrade in 1909, and by around 1912, he became interested in the mystery of the cause of the ice ages. A French mathematician, Joseph Adhemar, had pointed out in 1842 that the amount of heating of the Earth by the sun varies due to wobbles in the Earth's axial tilt. In the summer, the sun reaches a point that is vertically above the Tropic of Cancer so that the sun shines all day long at the North Pole and at the highest northern latitudes. Adhemar calculated that over a 22,000 year cycle, the tilt varies between 22.1° and 24.5° (it is presently 23.5°); the greater the tilt, the higher the sun would appear during summer at the poles, making the summers slightly warmer. Sometimes the Earth's tilt is smaller than it is now, so that less sunlight reaches the highest latitudes in summer. Adhemar proposed that during such times (small tilt with cooler polar summers and longer winters), the colder climate could trigger an ice age.

A Scottish geologist named James Croll expanded upon Adhemar's idea by including variations in the eccentricity of Earth's elliptical orbit. Due to gravitational interactions with the other planets, the Earth's orbit is sometimes more elliptical than at other times. An elliptical orbit means that the Earth is sometimes closer to the sun and sometimes further away. When winter in the northern hemisphere corresponds with a phase when Earth is furthest from the sun (and when the eccentricity of Earth's orbit is greatest), Croll surmised conditions were favorable for the start of an ice age. His ice age theory was summarized in his book *Climate and Time in Their Geological Relations* published in 1875. Although it was an interesting idea, Croll's hypothesis for the cause of the ice ages was not widely accepted by the scientific community due to lack of evidence.

Against this backdrop, Milanković began his mathematical investigations into the cause of the ice ages. He analyzed the main possible astronomical causes of changes in heating and cooling of the Earth, and he recalculated their periodicity. First is the eccentricity of Earth's elliptical orbit that Croll had worked on, which Milanković calculated has a 100,000-year cycle between maximum and minimum values. Also included was Adhemar's axial tilt theory, which Milanković calculated has a 41,000-year cycle (between 22.1° and 24.5°).

Lastly, he calculated the *precession* of the cycle of summer and winter seasons, which has a 23,000-year cycle. The precession of the seasons was first described by the ancient Greek astronomer, Hipparchus in 130 BC. If you think of the Earth circling the sun like the hours of a clock, sometimes summer (in the northern hemisphere) happens when the Earth is located at 12 noon. Over a period of 23,000 years, the occurrence of summer moves around the clock, from 1 PM, 2 PM, etc. In practical terms, it means that the position of the stars gradually changes so that the North Star, for example, which is presently Polaris, appears to move out of its position to be replaced by a different star.

Next, Milanković carried out a series of exhausting calculations, to estimate the amount of heat received at three different latitudes, 55°, 60°, and 65° north, for the past 650,000 years. The calculations were all done manually, of course, since computers had not been invented, and it is thought that it must have taken Milanković 100 days of nonstop computations to produce his findings. The result was a series of curves, showing the heating and cooling cycles due to the combined effects of all three astronomical factors. The curves first appeared in a book titled *Climates of the Geological Past*, published in 1924 by Wladimir Köppen and his son-in-law Alfred Wegener (of continental drift fame).

Importantly, Milanković concluded that it was the cooler summers (rather than colder winters) that caused the ice ages, contrary to the earlier theorists. Cool summers coincide with the combination of the sun being furthest away when it is summer in the northern hemisphere coinciding with smaller axial tilt. Cooler summers meant that snow that had fallen the previous winter would not all be melted. He showed that this situation had arisen about once every 100,000 years over the last 650,000 years. In other words, Milanković believed that there had been perhaps 20 or more ice ages in the past 2 million years (the Pleistocene).

Milanković prepared his findings in his masterpiece *Canon of Insolation of the Earth and Its Application to the Problem of the Ice Ages*. He sent his book away to be published in 1941, just as the chaos of World War 2 descended upon Europe and the rest of the world.

Milanković was a man ahead of his time. He was also a Serbian, and he wrote his book in German. His research had to be stopped for the duration of World War 2, and when the war ended in 1945, Yugoslavia was politically isolated from the western world as a satellite state of the Soviet Union. His book was not translated into English until 1960. Even then, there was evidence for only four ice ages in the Pleistocene, not 20. By the time he passed away in 1958, his theory for the cause of the ice ages had been dismissed by the scientific community and was mostly forgotten.

But not completely forgotten. Proof that Milanković was right came from an unexpected source – sediments from the bottom of the oceans.

***

Most people will have seen sedimentary rocks at some point in their lives. These are the layered rocks you might see along the side of the road, where excavators have made a road cut. Road excavations are the delight of geologists around the world. Perhaps you have seen groups of people standing along the side of the road someplace, staring with fascination and engrossed in earnest conversation, pointing enthusiastically at pebbles or layers of sediment. Perhaps you wondered what these strange people were doing on the side of the road?

Every geology student has spent some time on field excursions where their professor has explained the age and significance of the exposed layers of rock in a road cut. The professor will patiently explain the kind of environmental setting likely to have existed when the layers were first put in place, what was alive at the time inferred from fossils, and so on. These geological interpretations are often a kind of "geo-poetry" in the sense of Harry Hess and his "history of the oceans" paper; they are deductions and best estimates backed up with observations and some data. In many cases, the explanations given are essentially working hypotheses, waiting for further testing. Other times, there have been textbooks devoted to the origin of the rocks in question, and the story of how they were deposited and what they contain has been rigorously debated and discussed over many years.

Nearly all sedimentary rocks were originally formed under water, either by a river, in a lake, in an estuary, or in the ocean – some are volcanic ash beds, and a few wind-blown sediment deposits are also found, known as "aeolianites." Sedimentary layers represent a past event that occurred in a particular depositional setting. It may have been a single catastrophic event, like the eruption of a volcano or a mud slide, or it may have been a series of events like the steady migration of a sand dune across the seabed (daily or weekly layers). It may have been a change in the seasons such as melting glaciers or blooms of tiny plants and animals whose shells piled up into thick, annual layers. No matter how they formed, each layer is like the page of a book. The contents of each layer, the fossils and minerals, and the way the grains are organized record what was going on in that place at that time. Sedimentary rocks contain the history of the Earth, written in stone. These are the books that geologists read.

Over geologic timescales, sediments are buried, compressed, and turned into rocks. But we can also dig into these layers while they are still young and fresh, lying all soft and gooey on the seabed, to find out what was happening in the recent past, over the recent years, decades, or hundreds or even many thousands of years. This is a bit like crime scene investigation work, forensic science, to search for clues as to what was going on at a location, based on what was deposited within the different layers of sediment.

***

The breakthrough in Milanković's theory arrived in the 1970s, when the analysis of deep-sea sediment piston cores allowed a connection to be made between the past climate and the chemistry of the sediments. The most important of these technologies (apart from being able to collect the cores in the first place!) was the analysis of oxygen isotopes that we encountered earlier in relation to zircon crystals.

You may recall that there are two common oxygen isotopes: "light" oxygen-16 and "heavy" oxygen-18. When water evaporates from the ocean, more of the "light" oxygen-16 evaporates than "heavy" oxygen-18. Rain and snow contain more "light" oxygen-16 than "heavy" oxygen-18 in comparison with ocean water. When the world is in an ice age, there is a lot of freshwater containing "light" oxygen-16 locked up in continental ice sheets (enough to lower global sea level by 130 m). This means that the ocean contains more "heavy" oxygen-18 atoms during ice ages. When the ice sheets eventually melt, the "light" oxygen-16 is released back into the ocean, and the balance changes back to normal (if we agree that "normal" means when there is not an ice age).

Here lies the connection between climate and sediments: plankton grow their shells using calcium, and carbon dioxide dissolved in seawater to make calcium carbonate ($CaCO_3$, which contains oxygen). The calcium carbonate in plankton shells contains a mixture of both "light" oxygen-16 and "heavy" oxygen-18, with more or less oxygen-18 depending if there is an ice age or not. Sediment samples are taken from different layers, at intervals down a core, and the amounts of oxygen-18 and oxygen-16 in the plankton shells are measured (using a mass spectrometer). In this way, marine geologists built records going back in time and discovered that there were times when the oceans were enriched in "heavy" oxygen-18 atoms (indicating an ice age) and other times when the sediments were enriched in "light" oxygen-16 (indicating normal, interglacial conditions). The sediment core records showed when ice ages took place, how often they occurred, and how long they lasted.

And the results proved that Milanković was absolutely right!

Over the last 1 million years, ice ages have occurred once every 100,000 years, just as Milanković predicted. This finding was a result of the major research program called "Climate: Long range Investigation, Mapping, and Prediction" (CLIMAP) funded by the US National Research Foundation. The initial work, described in a benchmark paper published in 1976,[5] went back 450,000 years based on two core profiles collected in the southern Indian Ocean (Fig. 7.2). Later research findings extended the time series and expanded the spatial coverage around the globe.

Since the 1970s, hundreds of deep ocean sediment piston cores have been collected, and a time series of oxygen isotopes going back over the last 2.5 million years has been compiled. Ice cores have been drilled 2 to 3 kilometers into Greenland, and Antarctic ice sheets have also been analyzed. The longer records showed an interesting result. Ice ages occurred once every 100,000 years up until about 1 million years ago. But prior to that, they were even more frequent.

Between about one and two and a half million years ago, the frequency of ice ages was once every 40,000 years. This corresponds to the axial tilt frequency that Milanković had predicted. Scientists don't know why the frequency of ice ages

---

[5] Hays et al. (1976).

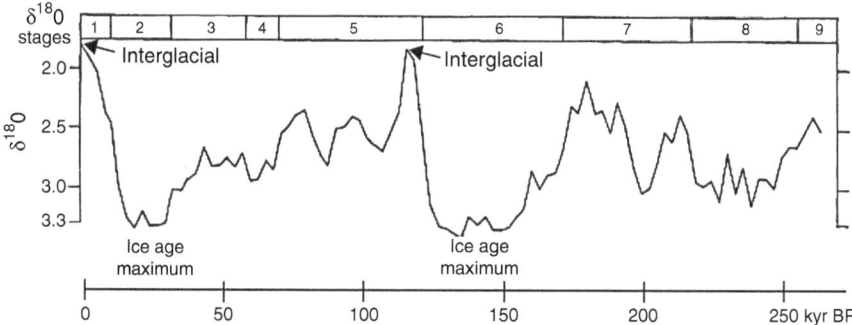

Fig. 7.2 Oxygen isotope curve from core RC-120 collected from a depth of 3135 m in the southern Indian Ocean (from Hays et al. 1976) showing variations in the ratio of oxygen isotopes 18 and 16 over the last 250,000 years

changed from 100,000 to 40,000 years around 1 million years ago. But it apparently did!

The Milanković theory for an astronomical cause of glacial cycles is now generally accepted as the most likely explanation for the *periodicity* of the Pleistocene ice ages. But there are questions left unanswered. Chief among them is the question of what was the initial cause of the first Pleistocene ice age two and a half million years ago? What was the trigger in the first place? No one knows for sure, but perhaps it was caused by something like the fluttering of a butterfly's wings.

***

Could the fluttering of a butterfly's wings in the Amazon jungle actually trigger a convoluted series of events leading to a hurricane in Florida? This is the "butterfly effect" metaphor from chaos theory, meaning that a small change in a system can be the trigger for greater consequences. If the "butterfly effect" seems remotely possible to you, then you should have no objection to the suggestion that the rising up of the Isthmus of Panama was a contributing factor, if not the direct cause, of the Pleistocene glaciation. It is complicated and not fully understood, but the basic idea[6] goes something like this:

Raising of the Isthmus of Panama blocked warm tropical water flowing through the so-called Central American Seaway. The blockage caused an intensification of the Gulf Stream which carried greater amounts of warm salty water into the North Atlantic. The warm salty water was (and still is) converted into cold salty bottom water to drive the Great Ocean Conveyor. But it also carried more moisture into the Arctic latitudes, which was transported over Europe and western Russia. This excess moisture increased the flow of rivers into the Arctic Ocean. Today, the Arctic actually receives 10% of global river discharge, even though it covers only 3.6% of global ocean area.

---

[6] Haug and Tiedemann (1998).

The excess freshwater input creates a layer of low-density water floating on the surface of the Arctic which stratifies the water column. When the Arctic became strongly stratified, the surface waters cooled down without mixing, making the surface of the ocean liable to freezing. The excess moisture therefore drove the expansion of sea ice. Meanwhile, on land, the excess moisture resulted in the formation of continental ice sheets. The lower albedo of snow and ice on land plus the floating sea ice on the Arctic Ocean, caused further cooling to create the conditions needed for Milanković's astronomical processes to take effect.

The Arctic Ocean is very isolated and nearly surrounded by land. The Bering Sea is very shallow with a sill depth of only around 45 m (150 feet) and so is not accessible for exchange of deep-water masses between the Arctic and North Pacific Oceans. Similarly, the Canadian archipelago west of Greenland is a maze of shallow continental shelf channels that eventually feed into the Labrador Sea, but they do not permit the exchange of deep water masses.

The only deep-water connection of the Arctic to the rest of the world ocean is via a narrow channel between the island of Svalbard and Greenland, known as the East Greenland Rift Basin. The width of the channel between shelf edges is a mere 290 km (180 miles), and although it is locally up to 5000 m (16,400 feet) deep, it has a sill depth of only 2500 m (8200 feet). This means that any water below that depth in the Arctic Ocean that is too dense to rise over the 2500 m (16,400 feet) deep sill is trapped in the basin. The isolation of the Arctic means that the coldest waters do not escape and a large amount of cold water (with a large thermal mass) is stored there. In this way, the Arctic can cool down and freeze over very easily.[7]

But this explanation is still not completely satisfactory. The problem is that the rising of the Isthmus of Panama blocking the Central American Seaway and diverting the Gulf Stream northward occurred around 4.2 million years ago, whereas the Pleistocene glaciations didn't get started until around 2.7 million years ago. All we can really say is that the rising of the Isthmus of Panama and diversion of the Gulf Stream were extra factors that pushed the Arctic closer to a tipping point so that anything that caused even a slight amount of extra cooling (be it sun spot cycles, volcanic eruptions, or whatever) could tip the balance. At present, we don't know the exact cause or trigger that began the Pleistocene glaciations 2.7 million years ago. It's another mystery.

What we do know is that the Earth's climate is incredibly sensitive to small changes that cause warming or cooling. The present-day configuration of continents, oceans, and the atmospheric system is poised on the threshold to jump from warm, interglacial phases and back into ice ages, by any changes in the processes that cause either warming or cooling. Milanković theory provides an explanation for the very small astronomical variations in solar insolation needed to push the Earth into ice ages or warmer (interglacial) warm periods. The difference between an ice age and the interglacial climate we enjoy at present is a mere few percentage points of greater or lesser solar insolation (heating). The average global temperature during the last ice age was only about 5 °C below present (preindustrial) levels.

---

[7] Tietsche et al. (2011).

The astronomical forcing is enhanced and reinforced by other processes, like "positive feedback." What do we mean by "positive feedback"? Take, for example, the scenario that Earth is descending into an ice age. Glaciers and ice sheets are expanding. But glaciers and ice sheets have a high albedo so, as they grow, more solar energy is reflected back into space, making the world colder. As ice sheets expand, forests with their low albedo shrink, decreasing the amount of absorbed sunlight, making the world colder still. These are examples of positive feedback driving the Earth into an ice age.

In the opposite scenario, the Earth is coming out of an ice age, and melting glaciers and ice sheets reveal land and vegetation that have a low albedo, increasing the amount of absorbed sunlight, making the world warmer. Open ocean has a lower albedo than floating sea ice, so as sea ice melts the amount of absorbed sunlight increases, making the world warmer still. These are examples of positive feedback driving the Earth out of an ice age.

The information we have on the Pleistocene ice age cycles indicates that the transition from warm conditions, such as we now have, into ice age conditions is a long and gradual process. A typical 100,000 year cycle includes around 80,000 years of gradual cooling, descending into an ice age in which the maximum extent of ice persists for around 10,000 years, followed by a rapid warming (interglacial) phase lasting around 10,000 years. The maximum cold and warm periods (the peaks and troughs of the curves) each last only around 10,000 years, on average. The warm phase we are currently living in has already lasted for over 10,000 years. We are living at a rare time at the top of an extended, warm-period peak.

The last glacial maximum occurred between around 25,000 to about 18,000 years ago. This is when the ice sheets covering Europe and North America were at their greatest extent. What were the oceans like at this time?

***

The first obvious difference between the present-day ocean and the ice age ocean is that sea level was lower by 130 m (400 feet). This is the amount of water that was transported from the oceans and onto the land to create the major ice sheets. These thick ice masses were the Scandinavian Ice Sheet that covered Northern Europe, the Laurentide Ice Sheet centered on Hudson's Bay that covered all of Eastern Canada, and the Northeastern United States and the Cordilleran Ice Sheet that draped over Western Canada and the Northwestern United States. These ice sheets were typically 3–4 km thick but reached up to 5 km in thickness in places. As they crept slowly, downslope toward the ocean, they transformed the landscape by carving deep valleys into the land surface. Puget Sound, the Great Lakes, and the Baltic Sea were all formed in this way. The ice sheets also deposited huge piles of rock and sediment rubble, the glacial moraines.

Since sea level was 130 m (400 feet) lower than present, the area of the oceans was less because much of the shallow continental shelf regions were exposed as dry land. In fact, the area of exposed continental shelves (excluding Antarctica's shelf which was still submerged because of its great depth) was probably around 30 million square kilometers (11.6 million square miles). So, the total area of ocean during

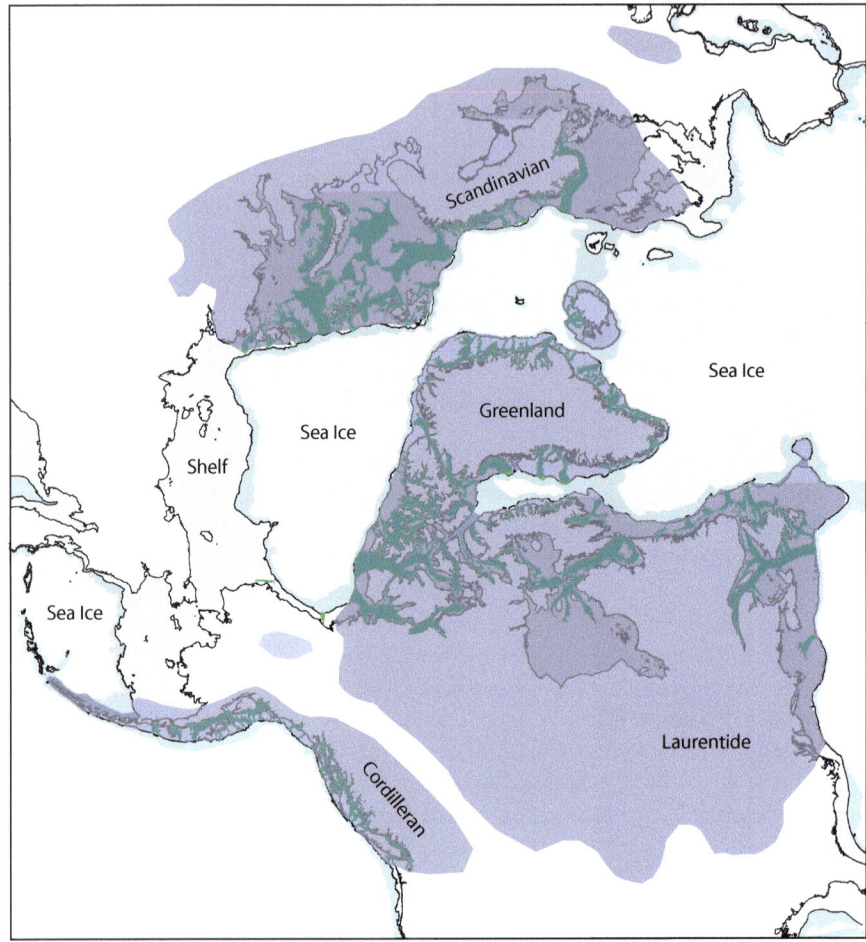

Fig. 7.3 Map of approximate locations of major ice sheets during the peak of the last ice age, around 21,000 years ago. The shelf area off Siberia was exposed land. Floating sea ice in the Arctic Basin may have been up to 1000 m thick, and sea ice extended as far south as the coast of France. Glacial troughs eroded into the shelf (shown in green) are from Harris et al. (2014)

the ice age was 30 million square kilometers smaller at around 330 million square kilometers (127 million square miles), compared with around 360 million square kilometers (139 million square miles) today.

A consequence of the exposure of the continental shelf was that, what are today shallow seas and island archipelagos were then extensions of the mainland, forming land bridges (Fig. 7.3). Perhaps the most famous is the Bering land bridge which joined Siberia and Alaska, when a vast shelf area was exposed covering around 1.6 million square kilometers (600,000 square miles). In Southeast Asia, the Indonesian islands including Borneo and Bali were joined to the Asian mainland. Tasmania and the island of Papua New Guinea were joined to Australia. The British Isles were joined to Europe (Brexit was not an option during the ice age!). These land bridges not only provided pathways for humans to migrate to all the continents on Earth (apart from Antarctica); they also blocked the flow of surface currents and had other major effects on the oceans.

At the time of the glacial maximum in the southern hemisphere, the Antarctic ice sheet advanced across much of the continental shelf. In South America, glaciers from the Andes extended across the shelf carving fjords and deep shelf valleys. Tierra del Fuego was joined to the mainland by a land bridge, and a thick ice sheet extended over the region. In Australia, a small ice sheet draped over the highlands of Tasmania which was joined to the mainland by a land bridge.

In the northern hemisphere, the continental shelf of Northern Canada is actually joined to northern Greenland (which is a part of the North American continental plate), and a massive ice sheet covered the archipelago, joining Greenland to Canada. In Northern Europe, ice sheets filled the Baltic Sea and smothered the Barents Sea, joining the island of Svalbard to the mainland.

In addition to the exposed shelf area were extensive floating ice shelves, essentially the floating continuation of continental ice sheets and glaciers. Examples of ice shelves exist today, and they occur along the East Antarctic margin and in the Ross Sea, in the Weddell Sea, and in Prydz Bay. These floating ice shelves are typically 1 km thick. At the point where they transition from a grounded glacier to a floating ice shelf with a cavity of seawater beneath, their bottoms drag along the seabed leaving grooves and piles of rocks and sediment that mark their passage.

During the last glacial maximum, ice shelves extended over all of the Antarctic shelf that wasn't already under a grounded ice sheet or glacier. The Arctic Ocean was reduced in area by around 50% because of the vast area of exposed continental shelf located offshore of Siberia. Evidence suggests that the upper 1 km of the entire Arctic Ocean was frozen solid and grounded on the seabed during the late Pleistocene (probably around 130,000 years ago), but it is not certain exactly when this may have occurred.[8] It may have also reached this thickness 20,000 years ago during the last glacial maximum.

Beyond the ice shelves is the sea ice zone, which grows and shrinks with the seasons. In the ice-age winter, the maximum extent of the sea ice zone around Antarctica extended close to a latitude of 45° south. Winter maximum sea ice in the northern hemisphere was located off the coast of France and extended across the North Atlantic to the coast of Maine. Sea ice extended well south of the Aleutian Islands in the North Pacific during winter. The Earth's floating ice shelves covered an area of as much as 15 million square kilometers, and sea ice covered another 15–20 million square kilometers during winter, leaving just 300 million square kilometers of liquid ocean surface during the ice age 20,000 years ago.

The exposed continental shelves, floating ice shelves, and sea ice zones had enormous consequences for ocean currents that were diverted into different pathways. Water flow from the Pacific to the Indian Ocean through the Indonesian archipelago was greatly reduced, and the Arctic Ocean was cut off from the Pacific by the Bering

---

[8] Jakobsson et al. (2016).

land bridge. Circulation in the Arctic Ocean was restricted to the only remaining, 290 km wide, opening between the islands of Svalbard and Greenland. The Gulf Stream flow was diverted south and did not reach north of the British Isles.

Today, the world's shelf seas are by far the most productive part of the ocean, accounting for over 90% of fisheries. In the ice age, the shelf area was greatly reduced, effectively deleting the most productive parts of the ocean, including the North Sea, the Bering Sea, and the large parts of the East Asian Seas. Furthermore, the ice age was a period of much lower rainfall, so river discharge was much less. This caused the flow of nutrients into the coastal seas to be lower, further reducing ocean and shelf productivity during the ice age. Consequently, there were probably fewer fish in the sea during the ice age compared with the warmer oceans we have today.

Species of plants and animals that occupy shelf habitats were squeezed into a narrow strip of shallow seabed that remained along the coast. Coral reefs, for example, occupied a much smaller area and retreated toward the equator as the waters at higher latitudes became too cold. The broad shelf areas around Indonesia and Australia, where corals grow at the present time, were exposed as dry land.

The great global ocean conveyor slowed down because the shelf seas, where bottom water is formed around the Antarctic, were covered by floating ice shelves. In the North Atlantic, the Gulf Stream did not penetrate as far north, and so it was also less effective in forming dense bottom water. The age of bottom water was about 50% greater, and the cycle of turnover was close to 1000 years in the Atlantic and perhaps 1500 years in the Pacific. Consequently, oxygen levels were reduced in the deep ocean, and the oceans contained a greater concentration of carbon dioxide (and were thus more acidic) than at present.

According to Milanković theory, the orbital pattern around 10,000 years ago brought the Earth closest to the sun during the northern hemisphere summer. This was the trigger for the end of the last ice age. Warmer summers melted the snow that fell the previous winter, and gradually, the ice sheets began to retreat. The melting ice brought about some amazing events in the oceans.

***

Unlike the gradual descent into ice ages, lasting 70 to 80, 000 years, ice ages come to an end much more quickly, taking less than 10,000 years ("quickly" is a relative term). The shrinking continental ice sheets melted away, pouring water back into the oceans. The Laurentide Ice Sheet that sat in the basin where Hudson's Bay is today was probably around 5 km (3 miles) thick. The ice sheet was so heavy that the land upon which it rested was depressed by as much as 1 km (3280 feet). It was so high in elevation that it deflected air masses and changed the Arctic climate, making it much colder and drier. Tongues of glacial ice extended from this ice sheet as far south as Iowa and Illinois. On the western side of North America was the Cordilleran Ice Sheet. It covered an area of over 2.5 million square kilometers (1 million square miles), including all of British Columbia, southern Yukon, southern Alaska, and the northern third of Montana, Idaho, and Washington states.

In Europe, the Scandinavian Ice Sheet filled the Baltic Sea with several kilometers thickness of ice and incised the seabed. The ice sheet covered all of Northern

Europe and extended over northern Russia, Germany, and Great Britain. It covered all of the shallow continental shelves of the North Sea, Barents Sea, and the Kara Sea in eastern Siberia.

Glaciers that reach from mountain tops down to sea level carry sediment and large rocks embedded in their icy bodies. Icebergs are born where glaciers terminate, and large pieces of glacier break off and float out to the sea. As the ice bergs melt, the rocks and sediment contained within are released, and they fall to the seabed. At the end of the ice age, the melting glaciers sent out "armadas" of icebergs. As they melted, the rain of rocks and sediment to the seafloor increased dramatically. The iceberg armadas and pulses of glacial, ice-rafted debris deposited on the seabed, are known as *Heinrich events*, named for the marine geologist who first described these deposits in 1988, Hartmut Heinrich.[9] A Heinrich event at around 16,800 years ago marked rapid melting, disintegration of ice sheets, and the first wave of icebergs at the end of the last ice age.

The melting and retreat of the Cordilleran Ice Sheet formed large meltwater lakes that were held back, behind dams of ice. The largest were Lake Columbia located along the border of British Columbia and eastern Washington, and Lake Missoula, located in Idaho and western Montana. The lakes filled with meltwater and repeatedly burst their dams between 15,000 and 13,000 years ago, causing the massive "Missoula Floods" across much of eastern Washington State. The floods left behind erosional features and giant sediment waves before cascading down the Columbia River and flowing out into the North Pacific Ocean (Fig. 7.4).

As the Laurentide Ice Sheet melted on the east side of North America, some of the meltwater became trapped in the depression that it created. The lake, known by scientists as Lake Agassiz (named for Louis Agassiz), filled the depression, and it grew larger and larger, covering much of what is today the Canadian provinces of Manitoba and Ontario. For a while, the water overflowed the lake to the south and emptied into the Gulf of Mexico via the Mississippi River. As the lake level rose and fell, it left a series of stranded beach deposits, like rings around a bathtub, which can still be traced today as low, scrub-covered ridges. The great ice sheet continued to melt, and Lake Agassiz retreated northward until, around 13,000 years ago, it passed the elevated glacial moraine (ridge) that separates the Mississippi River basin from the Great Lakes.

At this time, Lake Agassiz reached its greatest size. Trapped between the shrinking ice sheet to the north and a low ridge to the south, the lake filled to bursting point. Although we can't know exactly how it happened, at some location along the eastern side of the lake, there must have been a weak spot, a place where the mighty glacier was eroded by centuries of slow melting. And one day, 13,000 years ago, the ice dam bursts. In a single massive flood, the level of Lake Agassiz dropped 100 m (330 feet), and some 9500 cubic kilometers (2280 cubic miles) of freshwater cascaded over Niagara Falls and down the Saint Lawrence River into the North Atlantic Ocean. Did the flood last for weeks, months, or years? We may never know the exact details, but we do know that there were dire con-

---

[9] Heinrich (1988).

Fig. 7.4 Dry Falls, in Grand Coulee gorge in eastern Washington State USA. The cliff seen here was the scene of a massive waterfall during the Missoula Floods, between 13,000 and 15,000 years ago. (Photo by the author)

sequences for Earth's climate. The massive flood of ice-cold freshwater into the North Atlantic stopped the warming trend, and the world descended into an ice age that lasted another 1200 years.

The mass of freshwater discharged into the North Atlantic caused a reduction in bottom water formation and a slowing of the global ocean conveyor. This abrupt change in the routing of meltwater coincided with the onset of a worldwide cold spell known as the Younger Dryas, which lasted from 12,900 to 11,700 years ago. Global temperatures fell by over 2°; glaciers halted their retreat and advanced back down toward sea level. The climate became dryer in the northern hemisphere (the response of the southern hemisphere's climate is less certain but was probably less pronounced). A Heinrich event at around 12,000 years ago marked the end of the Younger Dryas ice age, as the ice sheets began a new phase of rapid melting.

There was one further hiccup for the global ocean conveyor that occurred around 8200 years ago. A meltwater lake located around the margins of Hudson's Bay burst through its ice dam and sent a pulse of freshwater into Hudson's Bay which made its way into the North Atlantic. The conveyor stalled, and temperatures in Greenland fell by 5 °C. The effect was a brief return to ice age conditions, with winter sea ice in the North Atlantic, lasting for 80 years.[10]

By around 6500 years ago, the Scandinavian, Laurentide, and Cordilleran Ice Sheets had mostly melted, global sea level stabilized near its present position, and the Earth entered its present warm, interglacial climate phase. Prior to this happening, the world's continental shelves were flooded by the rising sea levels. The changes in ocean tides and productivity that coincided with this event were dramatic. One of the most dramatic examples happened in the middle east, around Turkey.

***

While the melting of the Cordilleran and Laurentide Ice Sheets caused the great flooding events in North America, a different but equally amazing story was unfolding in another part of the world far away from the ice sheets. Although the Mediterranean Sea was not severed from the North Atlantic at the Strait of Gibraltar (the depth of the sill is 280 m), the Black Sea was cut off from the rest of the Mediterranean at the Bosporus Strait, which was a land bridge during the ice age. The Bosporus Strait, in Turkey, is today a narrow channel only 700 m wide and 13 m deep at its most shallow point. It is also the only connection between the Black Sea and the Mediterranean Sea and to the rest of the world ocean.

During the ice age and until around 11,000 years ago, the Bosporus Strait was exposed as dry land, and the Black Sea was cut off from the Mediterranean Sea. The Black Sea was essentially a big lake, and its level fell, possibly by as much as 100 m below its present position because of the arid climate. The rate of river inflow was less than the evaporation rate so the Black Sea "lake" shrunk to a smaller size. Evidence for this lowering is provided by discoveries of buried river channels and shorelines at great depth along the margins of the Black Sea.[11]

Rapid melting of the Eurasian ice sheet and the Alpine ice dome around 10,000 years ago resulted in meltwater flowing into the Caspian Sea which overflowed into the Black Sea until it, in turn, overflowed over the Bosporus and into the Mediterranean. The Black Sea became a giant, freshwater lake. But as the climate warmed and the meltwaters dwindled, the level of the Black Sea "lake" fell again. By 8500 years ago, the Black Sea had fallen to a level around 85 m (280 feet) below its present position, and the scene was set for a catastrophic flood event that would go down in (biblical) mythology – Noah's flood![12]

Evidence from carbon-14 dates of wood and shells in sediment cores taken from the seabed of the Black Sea show that by 8400 years ago, the Black Sea freshwater "lake" had become saline. In other words, the Bosporus land bridge had been breached allowing Mediterranean Sea water to flow into the Black Sea sometime before 8400 years ago. There is only a short, 100-year window of time unaccounted for between the existence of the Black Sea "lake" 8500 years ago with its surface at 85 m (280 feet) and the modern, salty Black Sea, with its surface close to where it

[10] Barber et al. (1999).

[11] Ryan et al. (2003).

[12] Ryan and Pitman (1998).

is today by 8400 years ago. The question is was the breaching of the Bosporus land bridge a catastrophic event, with a rapidly flowing seawater cascade? Or was it a kind of fizzler of a flood, with Mediterranean Sea water leaking slowly into the Black Sea over a period of 100 years? Research continues, and the question remains unanswered. Right now, only Noah knows for sure.

One interesting aspect of this story is that it has parallels with other locations around the world where a basin was isolated by a lower, ice-age sea level and then flooded by rising sea level in the early Holocene. The Gulf of Carpentaria and Bass Strait in Australia are two examples of shelf-perched basins where large brackish lakes existed during the ice age. Another is the Persian Gulf, which was also above sea level during the ice age.

The Gulf of Carpentaria covers an area of over 500,000 square kilometers (193,000 square miles) which was exposed during the last ice age, mainly as a dry arid landscape. In the center of the Gulf, there is a bathymetric depression that was transformed into a lake when the sill between the Arafura Sea and the Gulf was emergent and during some wetter climatic phases. At its greatest extent, the lake may have been as much as 500 km (310 miles) long, 250 km (155 miles) wide, and 15 m (50 feet) deep.[13] The ice age climate was mainly a dry period in Australia, and there were long time spans when the lake was either very small or absent (dried up).

In Bass Strait, located between the southern coast of Australia and the island of Tasmania, there is another large area of continental shelf of around 66,000 square kilometers (25,000 square miles) containing a central depression with a raised rim around it. The story here is similar to the one in the Gulf of Carpentaria; a large shallow lake is believed to have occupied a depression in the middle of Bass Strait during the ice age. In both cases, it is likely that Australian Aboriginal peoples occupied the lake shore to access fish, drinking water, and wildfowl resources.

The Persian Gulf covers an area of around 235,000 square kilometers (90,000 square miles), and its shallow depths are less than 50 m (160 feet) in most places. During the peak of the last ice age 20,000 years ago, what is today the Persian Gulf would have been a broad valley with an elongate lake at its southern end, close to what is now the central Gulf coastline of southern Iran. The lake would have received waters from the Tigris, Euphrates, Karun, and Wadi Batin rivers, and along the lake shore, anthropologists suggest humans occupied a "Gulf Oasis" around that time.[14] Rising sea level breached the Gulf sill depth of 80 m (260 feet) by around 12,000 years ago, after which the lake would have been transformed into a large estuary. The "Gulf Oasis" was submerged under the sea by 10,000 years ago.

The Persian "Gulf Oasis" and Australian lakes were not dramatically below sea level, and so the "flooding," when it occurred, was probably not the sensational event like the one that may have occurred in the Black Sea. But a lot of low-gradient land going under water in a relatively short space of time (i.e., within living memory of human occupants) is the likely explanation for the over 200 flood myths that exist

---

[13] Jones and Torgersen (1988).

[14] Rose (2010).

from cultures all around the world, including China, India, Europe, Africa, and South America.

***

Tidal rise and fall along the world's coastline is controlled to a large extent by the depth, width, shape, and orientation of the continental shelf. As the ocean tide approaches the coast from the deep ocean basins, it first encounters the shallow shelf edge. The shoaling depth causes the tide to increase in height. Next, the shape of the coastline can amplify the tide. The funnel shape of the Bristol Channel (Great Britain) coastline, for example, explains the 14 m (46 feet) tidal range that occurs there.

If the distance between the shelf edge and the coast is close to a harmonic of the tidal wavelength, the tidal range is increased even further. The largest tidal range that occurs today, 15 m (49 feet) in Canada's Bay of Fundy, is related to tidal resonance (as is also the case in the Bristol Channel). The distance from the shelf edge to the head of the Bay is almost exactly one-quarter of the tidal wavelength, and so the Bay of Fundy is perfectly tuned to the oceanic tide in the North Atlantic.

What Were Tides Like when Sea Level Was Lower During the Ice Age?

As sea level rose, the continental shelf was gradually flooded, and the tidal regime along the coast changed. In some places, the tidal rise and fall went through incredible transformations, from a small (microtidal) range of less than 2 m (6 feet) increasing to large (macrotidal) regimes greater than 10 m (30 feet) and back again. The reason this happened is because flooding changed the coastline shape, causing it to go in and out of tidal resonance with the ocean tide.

Computer models can accurately predict tides along today's coastline, but they can also be used to estimate what tides were like over the last 10,000 years, as the ice age ended. Models show that the Bay of Fundy did not attain its world record tidal range until around 5000 years ago. At the same time, tides along the Atlantic coast of the United States were nearly twice as high as they are today.[15] As the ice sheets melted and sea level rose, the tidal range grew bigger in the Bay of Fundy and smaller along the US Atlantic coast. In the Bristol Channel, the tidal range was large around 10,000 years ago and decreased around 5000 years ago before increasing again to its present levels. Computer models can also predict what tides will be like in the future – even a 1 m rise in sea level is predicted to cause changes in Chesapeake Bay tides of up to 10% higher than present.

***

Melting of the northern hemisphere's continental ice sheets caused flooding of the worlds continental shelves, creating approximately 29 million square kilometers (11 million square miles) of shallow shelf seas, an area greater than the continent of North America. Such shallow shelf seas were comparatively tiny during the ice age, practically nonexistent. Only the Antarctic shelf was fully submerged during the ice age, and it was completely capped by a floating ice shelf.

---

[15] Hill et al. (2011)

All of the plants and animals living on the shelf today are recent colonists that arrived in the last 10,000 years or so. Flooding and colonization of the shelf seas is a process that fundamentally transformed the oceans. The role of the shelf seas as the lungs of the ocean is critical (Chap. 6), but the shelf seas today are also the most productive parts of the ocean, accounting for over 90% of commercial fisheries and over 50% of primary production; in other words, the shallowest 10% of the ocean produces 90% of the fish.

All of the world's coral reefs have also colonized the shelf in the last 10,000 years. This includes Australia's Great Barrier Reef, which has been the focus of study and research on coral reefs for nearly two centuries.

The history of coral reef study on the Great Barrier Reef shelf dates back to the early to mid- 1800s, and the work of Charles Darwin, whose subsidence theory of atoll formation, caused much interest and debate on the general subject of coral reef evolution (although the Great Barrier Reef itself featured rarely in these debates). Some early workers extrapolated the evolution of atolls to the continental margins, evoking continental subsidence. Some thought the Great Barrier Reef was of great age and thickness, while others considered it to be thin and of Pleistocene or younger age.

The various theories were finally put to the test when, in 1926, the Great Barrier Reef Committee sunk a borehole on Michaelmas Cay. The hole penetrated 145 m (476 feet) through carbonate reef material and bottomed at 183 m (600 feet) in shelly quartz sands with glauconite and foraminifers. The bore did not retain any sample between 12.8 and 27.5 m (42 and 90 feet), but a difference in lithology was noted in the sections above and below this gap; solution and cementation were evident in limestones below compared with unaltered aragonite and high-magnesium calcite above. Other boreholes were collected at Heron Island in 1937 and later at Anchor Cay, Wreck Reef plus Capricorn, and Aquarius wells in Capricorn Channel. These results demonstrated the great thickness of limestone comprising the reefs, but in the era before radiocarbon dating, the age of the coral deposits was uncertain.

In the 1960s, prior to our current knowledge of plate tectonics and the cause and complexity of the Pleistocene glaciations, the accepted line of thinking was that all of the limestone (i.e., to a depth of 145 m or 476 feet or so) was of Holocene age (less than about 10,000 years old). The assumed thickness of the Holocene reefs required rapid and extensive upward growth. It also meant that present coral reef morphology must be attributed to reef growth under modern environmental controls.

Such was the underlying assumption to W.G. H. Maxwell's classification schemes of reefs.[16] Maxwell was one of Australia's leading reef experts in the 1960s, and his book *Atlas of the Great Barrier Reef* was the foremost authority on reef growth and geomorphology. Reef shapes were attributed to modern processes – some were deltaic shaped because they had grown on top of tidal deltas, others were more linear like barrier islands because of their supposed foundations, and so on.

---

[16] Maxwell (1968).

In the early 1970s, the proposed Holocene age of the Great Barrier Reef came into question from two independent investigations. First, in Belize (Central America), coral reef investigations discovered a relatively thin veneer of Holocene reef growth on top of what was found to be much older, recrystallized, and eroded Pleistocene reef limestone. New coral reefs had simply grown on top of long dead reefs that had been exposed on dry land during Pleistocene times of lower sea level. The erosion by dissolving away the old limestone by rain and standing water is a process known as "Karst," and it makes limestone caves, sinkholes, and stalactites.

The results from Belize inspired Prof. Peter Davies who was at that time employed by the Australian Bureau of Mineral Resources (presently Geoscience Australia), to reinterpret the 1937 Heron Island drill core. Davies proposed the existence of a Holocene/Pleistocene solution unconformity at about 20 m (66 feet) depth down the core, in which the age of the material directly above the unconformity was less than 10,000 years, whereas the underlying, weathered limestone was of much older, Pleistocene age.[17]

Peter Davies' new interpretation of the origin of the Great Barrier Reef's foundations proved to be a revelation to the marine science community. Throughout the late 1970s and 1980s, numerous additional bores were obtained from individual reefs on the Great Barrier Reef shelf, and accurate radiocarbon dating has proven that the Holocene reef growth was initiated by 8000–9000 years ago and that the thickness of Holocene reefs is typically 10–15 m (30–50 feet) in the northern and southern sections of the Great Barrier Reef and somewhat thicker (20–25 m or 66–82 feet) in the central Great Barrier Reef.[18]

The geomorphology of the Great Barrier Reef shelf is best explained in terms of erosional processes (including karst and fluvial erosion) during low sea level phases together with shelf-edge and platform reef growth processes during high sea level, interglacial periods. With every ice age, the Great Barrier Reef dies off and is eroded. This is followed during high sea level, interglacial phases, by new reef growth on top of the remains of the old reef. Reef growth, death, and regrowth have occurred over and over again, throughout the quaternary (last 2 million years) and further back in geologic time.

Indeed, the Great Barrier Reef is truly ancient. Sea level change is only one process that has affected its growth and evolution. Plate tectonics has also played a critical role. After splitting off from Gondwana around 80 million years ago, the Australian continent has slowly moved northward, carrying the Great Barrier Reef along with it from temperate into tropical waters. This explains why the reef deposits are much thicker (over 2000 m or 6560 feet) and older (around 20 million years) in the Gulf of Papua at the Reef's northern end and much thinner and younger in the south.[19]

***

---

[17] Davies (1974).

[18] Holpley et al. (2007).

[19] Davies et al. (1987).

And now we have arrived in the Holocene epoch, which began around 10,000 years ago. The ice age has ended, sea level has stabilized around its present position after around 6500 years ago, and the world's climate has enjoyed a lengthy, warm, "interglacial" period. All of human recorded history and civilization has occurred in the Holocene, which has given us a biased view of global climate (Hasn't it always been like this?). Our experience of what climates are even possible on Earth is based on a tiny fragment of time, equal to 2.7 mm of the 1.1 km geological time walk outside the Geoscience Australia building.

It is time now to explore our ocean world as it exists today. We will visit different submarine environments and habitats that have only recently been discovered and explored. Some of these are familiar places: estuaries and deltas, coral reefs, and kelp forests. Other places will be less familiar: canyons and trenches, hydrothermal vent communities that live on sulfur-reducing bacteria, black corals that live on the sides of seamounts, fracture zones, and sunken continents.

We shall explore the 70% of Earth's surface that comprise our mysterious oceans. The problem we face is that we cannot see the ocean floor because the oceans are opaque. But we can use our imagination and transform water into a transparent substance. Light will reach to the ocean floor for the first time in Earth's history, and we will be the first humans to see these features in broad daylight! So, let's get started!

Chapter 8
The Continental Shelf

*"Fringing-reefs are thus converted into barrier-reefs; and
barrier-reefs, when encircling islands, are thus converted into
atolls, the instant the last pinnacle of land sinks beneath the
surface of the ocean."*
Charles Darwin
The structure and distribution of coral reefs, 1842.

Abstract In the next two chapters, we will take a tour aboard an imaginary hover
car to visit the world's continental shelves and the deep ocean. The continental shelf
(as the name suggests) is the submerged part of the continents, and the features we
see underwater are comparable to those we see today along the coast. The continental shelf is where rivers disgorge their loads of sediment and where glaciers built
moraines during the ice age, providing the material needed to make sandbanks.
Sand is in fact the second most used commodity on Earth after freshwater. It is a key
ingredient of concrete and is also used to replenish beaches that are eroding because
of rising sea level. In this chapter we will observe the impacts of bottom trawl fishing on seabed habitats and consider the need for marine parks. We will explore hidden coral reefs in the Gulf of Carpentaria and see how bleaching of corals is
happening more and more often. Has a tipping point been passed? Will we be able
to save the coral reefs? To avoid burning fossil fuels, we should look at harnessing
the ocean's tide power as a renewable power source. Human impacts on land often
reach the ocean. The Ok Tedi gold mine disaster in the Fly River, Papua New
Guinea, is an example of this.

Keywords Algae line · Pipette analysis · Sand mining · Moreton Bay ·
Lag deposit · Sediment sorting · Kelp forest · Ghost fishing · Bottom trawling ·
Microplastic · Gulf of Carpentaria · Mesophotic coral reef · Coral bleaching ·
Ocean warming lag effect · Tipping point · Tide power · Mike Collins ·
Side scan sonar · Bristol Channel · Fly River Delta · Ok Tedi mine · Clinoforms ·
Coral Sea Current

If we drained away all the water from the oceans, what features, what creatures, and what landscapes would be revealed? Let's imagine that the oceans have become transparent today; all the fish, plants, and animals are safely "floating" in imaginary water that is totally transparent and that we can walk through and breathe, just as we can walk through the air. Maybe walking is not a good idea, since we have a long way to travel. We'll need to drive in a fast submarine hover car to take this tour (wheels are likely to get stuck in the mud, so we'll take a hover car instead). Let's fly across this foreign landscape revealed for the first time to human eyes. This quick tour will be an introduction to seafloor features that we will visit later for a closer look.

The place where land ends and ocean begins, the coastline, is transitory in space and time. Sea levels rise and fall, causing the coast to move landward and seaward. Waves are eroding it and causing it to recede in places. Elsewhere, rivers disgorge their loads of sediments into estuaries and deltas, and the coast advances seaward. These are exciting places for geologists being such dynamic environments where one can conduct experiments to work out how sediment layers and beds were laid down in ancient rocks. We'll need to come back to the coastline later for a closer look.

Leaving the shore behind us, we find ourselves flying across a more or less level terrace for about 60 km (35 miles).[1] This terrace is the continental shelf, and its width varies from over 400 km (250 miles) in the Arctic Ocean and Antarctica to an average of less than 10 km (6 miles) along some continental margins that are active plate tectonic margins (adjacent to subduction zones). In some places, there is no continental shelf at all, and we see only a steep slope dipping away from the coast.

The features we see on continental shelves are quite familiar to us because they look very much like features we have seen on land along the adjacent coastline. This is no surprise because the continental shelves are extensions of the continents. They were mostly exposed during the last ice age, 18,000 years ago, when sea level was lowered by around 130 m (427 feet) and many features we see relate to this time. There are river valleys, hills, rocky knolls, and flat-topped plateaus. Currents have piled up large sand dunes and ridges that look and behave a lot like the ones we can see today along the coast and further inland in the desert.

The rocks and sediments close to shore have a greenish tinge to them from the algae coating the seabed. But as we have moved into steadily deeper water, the green has faded. This is a sort of reverse of the "tree line" seen on the side of mountains, where the higher you go, the fewer trees you see until at some elevation there are no trees. In the ocean, there is a depth below which sunlight is too weak to support benthic algae, so we get an "algae line" instead of a tree line. The algae line varies from place to place depending upon latitude (in high latitudes the angle of the sun means less depth of sunlight penetration) and water transparency (depending mostly on how much plankton and sediment is suspended in the water column). The algae line will occur at around 30 m (100 feet) depth in most places, but its depth varies widely from place to place.

---

[1] The average width of the continental shelf is 57 km: Harris and MacMillan-Lawler (2016).

Once we have crossed the continental shelf, we will arrive at the shelf edge. This remarkable feature has no parallel on the Earth's surface above sea level – the view from here, looking out over the ocean basin, is unlike anything humans have ever seen. Looking over the edge of the Grand Canyon may give you some idea, but there is no opposite canyon wall to see from the shelf edge. From the shelf edge, the sea-floor panorama extends to the distant horizon (Fig. 8.1).

Park the hover car, and step outside onto an "Atlantic"-type margin, and you are standing on the crest of a slope that dips away into the distance until, 40 or 50 km

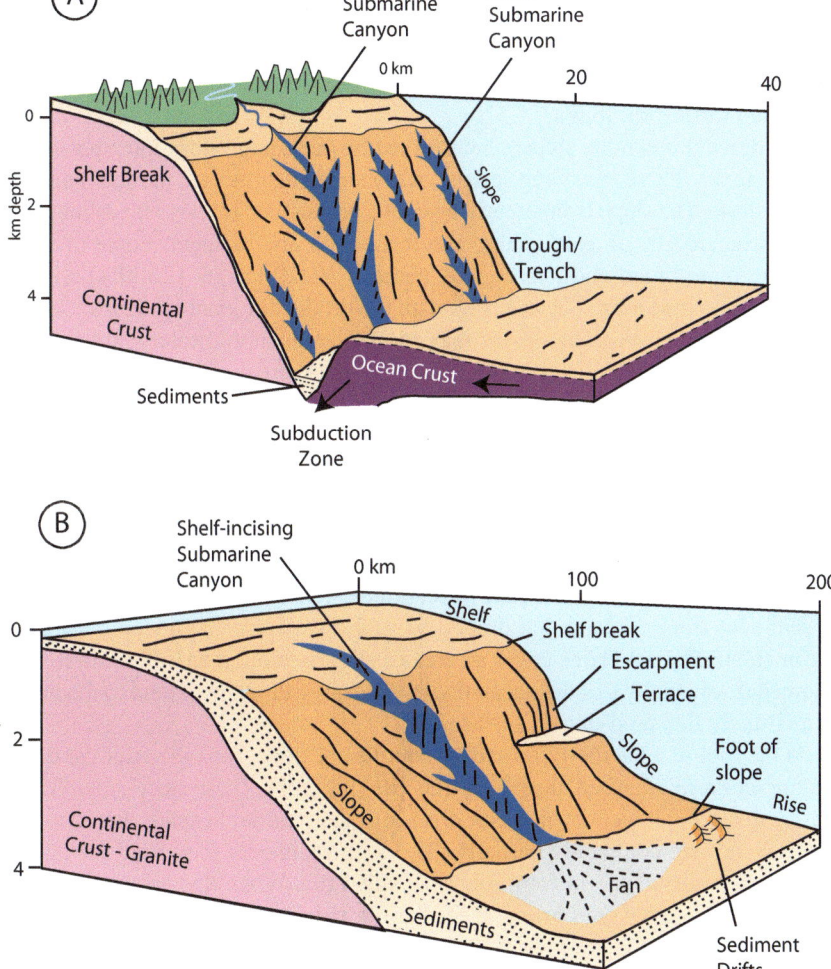

Fig. 8.1 Three-dimensional drawings showing: (**a**) active, Pacific-type continental margin with narrow shelf and subduction zone; (**b**) passive, Atlantic-type continental margin with broad shelf and thick sediments forming a continental rise adjacent to the margin

(25 or 30 miles) away, you may see the ground level out onto the continental rise around 4 km (2 miles) below. If, however, you parked the hover car on a "Pacific" type margin, you will see a slope that dips steeply away into the distance until 30 or 40 km (20 or 25 miles) away the ground drops out of sight into a deep ocean trench. The trench is over 6 km (3.5 miles) deep, and the bottom is too far away to see from where we are standing. The far side of the trench is over the horizon and beyond our range of view. Beyond the trench, the seabed rises to a depth of around 4 km (2.5 miles).

Back in the hover car, we begin to descend from the shelf edge down the incline of the continental slope: gentle and muddy in some places, steep and rocky in others. The sunshine seems to be getting brighter. That's because the amount of phytoplankton in the water column drops off dramatically as we leave the productive waters of the continental shelf. There are patches of greenish "clouds" in the ocean above us, where eddies have spiraled parcels of shelf water off the shelf, over the slope, and further out to sea.

Our descent down the slope is interrupted by occasional steps interspersed with steeper slopes. These are slump scarps, created by instabilities in the sediment draping the slope. The slope is inherently unstable, and sudden massive landslides, commonly triggered by an earthquake, may happen at any moment. Some of the flat steps are quite broad and terrace-like, perhaps 10–20 km (6–12 miles) wide and 100 km (62 miles) long. These terraces are the product of giant landslips.

An area of concentrated slumps may form a large amphitheater that has incised into the slope for some distance. The largest of these erosional features may evolve into submarine canyons, in which steep escarpments, with near-vertical walls and overhanging cliff faces, mark the eroding heads of the canyons. Continuous erosion of the canyon may eventually cause it to cut into the shelf to create some of the most spectacular features of the ocean floor that we shall explore later.

At the foot of an Atlantic-type (passive) continental slope is the beginning of the flattest feature on Earth's surface, the abyssal plain. These are the smooth, sediment-draped seabed areas located next to the continents. They are created by the shedding of sediment from the land that drapes over (and smooths) the underlying basaltic ocean crust. Abyssal plains cover an area of slightly more than 100 million square kilometers (about 28% of the ocean floor).[2] The abyssal plains extend to the horizon as a virtually flat, mud-draped surface.

At the foot of a Pacific-type (active) continental slope, we encounter the deepest feature on Earth's surface, the ocean trenches (Fig. 8.1). These steep-sided features are created by subduction of ocean crust into the mantle, beneath the continents. The trenches are commonly partially filled with sediments shed from the adjacent continents and slumped in from the steep sidewalls. When this happens, the sediment fill creates a flat bottom, and these flat-bottomed trenches are called "troughs." Trenches and troughs are rare in the ocean, covering less than about 1.3% of the ocean floor.

---

[2]All statistics on the areas and numbers of seabed geomorphic features are taken from Harris et al. (2014).

As we progress away from land and toward the mid-ocean ridge, the seafloor appears to become increasingly bumpy. The flat, sediment-draped surface of the abyssal plain is blemished by elongate rocky ridges and volcanic cones that pierce through the sediments from below. Volcanos and ridges become more and more common as we progress away from land and nearer to the mid-ocean ridge. Volcanic peaks that attain an elevation exceeding 1 km (3280 feet) above the level of surrounding seabed are called "seamounts." There are around 10,000 seamounts in the world ocean that we shall explore later.

As we continue our hover car journey away from the land, the sediment drape becomes much thinner, and there are many small hills and some elongate valleys. This is the realm of "abyssal hills" that are 300 m (980 feet) to less than 1000 m (3280 feet) in height and that extend between the mid-ocean ridge and the flat abyssal plains. Covering 150 million square kilometers (58 million square miles), abyssal hills are the largest category by area, of ocean floor type.

Finally, we reach the mighty mid-ocean ridges with their lava-filled rift valleys. There is a gradual rise from the abyssal hills to the summit of the mid-ocean ridges which are over 1 km (3280 feet) above the level of surrounding seafloor. Mid-ocean ridges cover an area of around 8 million square kilometers (3 million square miles). They are a fairly narrow belt that wraps around the globe like the seams of a baseball.

Extending away from the mid-ocean ridge at right angles are long and narrow fracture zones. These elongate, linear features are also huge. In the North and South Pacific, fracture zones extend across the seabed over distances exceeding 3000 km (1864 miles). Up close, a fracture zone looks like a series of three or four ridges, 500–1000 m (1640–3280 feet) in elevation above the surrounding seafloor, with deep valleys in between. The ridges are made from broken pieces of ocean crust that have buckled and been thrust upward.

Over the next two chapters, we will explore these features one by one to learn their secrets. Starting from the coast, we will make our way into deeper waters, exploring different marine habitats and the kinds of creatures that live upon them. And we shall learn how human activities have changed them in recent years.

***

Continental shelves are the submerged margins of the continents that happen to lie below sea level at the present time. Shelf seas cover an area of around 32 million square kilometers and are undoubtedly the best known part of the ocean being close to the coast and port cities where people live. They provide over 90% of the fisheries and around one-third of global petroleum production. The continental shelves were largely exposed during the ice ages when global sea level was about 120 m (360 feet) lower than it is today. This explains why features that appear on continental shelves are often comparable to those seen on the adjacent coast.

Over the last 50 years, mapping of the world's continental margins has revealed the complex character of continental shelves. There are some shelves that have been buried in thick sediment layers that contain valuable oil and gas deposits. These thick deposits extend in places out into the ocean basins, for example, the Gulf of Mexico and eastern United States; the shelves of Brazil, Argentina, Western Europe; and along parts of the eastern margin of Africa.

In some locations the shelf is narrow, rocky, and sediment-starved, for example, the coasts of Chile, eastern Australia, and California. Elsewhere, glaciers have eroded valleys and troughs, deeply incised into the shelves of Antarctica, Northern Canada, southern Chile, and Norway. At some places in the tropics, coral reef growth has built a shelf edge barrier (think of the Great Barrier Reef), which has been backfilled with sediment to create a characteristic shelf profile having an outer "rim." Finally, there are places where there is essentially no shelf at all, where the water depths drop away hundreds of meters within a kilometer of the beach. This is the case, for example, along the coast of the Philippines, the southern coast of Saudi Arabia, and the southern coasts of islands in the Indonesian archipelago.

Our exploration of the continental shelves has revealed their complex geologic history, which differs from place to place. The continental shelves are the most important parts of the oceans for humans, socially and economically. But we still have much to learn about how they evolved and how their ecosystems function.

***

Continental ice sheets had a great influence on the terrestrial landscape, but we forget that about 20% of the world's continental shelf was also glaciated during the last ice age. The continental shelves that experienced glaciation were obviously in the Arctic and Antarctic, but those along the Pacific coast of North America from Alaska as far south as Washington State, on the US east coast as far south as Manhattan, southern Chile, New Zealand, and the North Sea of Western Europe, were also glaciated at the peak of the last ice age.

During the ice age, glaciers flowed seaward from the elevated interior and cut deep furrows into the shelf sediments, leaving 1 km (3000 feet) deep scars, called glacial troughs, across the continental shelf in many places. Puget Sound, near Seattle, is the southern end of a glacial trough carved by the Cordilleran Ice Sheet 20,000 years ago.

In the late 1970s, students enrolled in the undergraduate oceanography program at the University of Washington in Seattle were expected to complete a field project in their senior year that involved doing a small piece of research work using the Universities' small boats and laboratory facilities. I was inspired by my professors, especially Dick Sternberg and James Dungan-Smith, to seek a project that applied principles of oceanography and geology to understand how the seafloor had evolved over geologic time.

The project that I worked on with my college buddy, Dave Collins, was based on a scale model of Puget Sound made from plywood and fiberglass and painted pleasant blue and green colors. The model was filled with water, and its purpose was to simulate tidal currents flowing in and out of the Sound by means of a mechanical plunger that moved up and down to represent the tides. As the plunger went down, water was forced into the complex of bays and inlets of Puget Sound on the flood tide. When the plunger went up, water flowed back out to "sea" on the ebb tide. A movable dye injector allowed you to see which way and how fast the water moved at any location during the tide cycle.

In the 1960s the model was used to produce a book of tidal currents published by the University called *Tide Prints*. Physical models like this were at the peak of technology in the 1960s and 1970s but were later superseded by computer models. By 1980 the Puget Sound model had fallen into disuse and was by then languishing in the basement of the Harris Hydraulics Laboratory located next to the UW Oceanography Department.

Professor Alyn Duxbury allowed Dave and I to use the model to design our senior year project; we noticed that a jet of water flowing into Dyes Inlet, near Bremerton, formed a large gyre in the middle of the bay on each flood tide. We wondered if sediment swept into the bay would settle out with the finest clay particles in the middle of the gyre (where currents are the weakest) with coarser sands and silts surrounding the perimeter of the gyre. We figured that, over time, sediment would pile up in the middle of the gyre and form a mound of mud, and sure enough, there is a large mound on the seabed in the middle of Dyes Inlet.

The geomorphology of Puget Sound is the product of repeated glaciation over the last 2 million years (the Pleistocene Epoch). At the peak of the last ice age 20,000 years ago, a 1800 m (6000 feet) thick ice sheet filled the main basins of Puget Sound, sculpting the bedrock and leaving thick deposits of sediment (glacial moraines) along its path. After the ice retreated, sediment was washed into the Sound from rivers, causing mud and sand to partially fill in the deepest submarine valleys. Today water depths are typically over 200 m (650 feet) along the main basins of Puget Sound and up to 280 m (920 feet) at the deepest location. Strong tidal currents have subsequently scoured the floors of narrow passages of the finest sediment and transported it into the bays and the main parts of the Sound where currents are weaker.

How much of the present-day shape of Puget Sound is due to glacial processes, and how much can be attributed to postglacial rearrangement of the seabed by tidal currents and river sediment deposition? Was the mound on the Dyes Inlet seabed of glacial origin or the product of modern oceanographic processes? This question was at the heart of the student project that Dave and I had decided to undertake.

On a drizzly gray day in 1980, Dave and I sailed with Prof. Duxbury on the RV *Onar* from the UW campus across Puget Sound. We cruised past the USS *Missouri*, mothballed at that time in the Bremerton Naval Shipyard and into Dyes Inlet. To collect the sediment samples we would need, Dave and I used a van Veen grab, a device designed in the early twentieth century by Dutch marine scientist, Johan van Veen. This grab is essentially a pair of steel jaws that are lowered on a wire to the seabed. On reaching the bottom, the jaws close to "grab" a bucket of surface sediment (called a grab sample). Over the next 2 days, we collected 40 grab samples, 1 from each station spaced about 500 m (1640 feet) apart in a grid pattern across the mud mound lying at the tidal gyre's center in Dyes Inlet.

We brought our samples back to Seattle and commenced the laboratory phase of our study, to measure the size of sediment particles. To do this we decided to conduct a pipette analysis on each sample. This technique is based on the fact that particles of a known density will fall through a fluid according to their size, with smaller particles falling slower than larger ones. The analysis starts with each sample being placed in a graduated glass cylinder mixed with water and shaken to thoroughly mix the muddy water. The cylinder is then placed on the lab bench, and a sample is drawn by pipette to measure the mass of suspended sediment.

As the mud settles toward the bottom, samples are repeatedly drawn by pipette at a fixed distance from the surface at predetermined time intervals. From physics, we can calculate how long it will take for a particle of known size to fall through the fluid, and then we can work out what size of particles are missing from the upper part of the muddy water mixture after a certain time has elapsed. For example, after 15 minutes we know that particles coarser than 16 microns will have fallen 20.8 cm. The difference in weight between a sample taken at 20.8 cm depth immediately after stirring and 15 minutes later is therefore the weight of sediment coarser than 16 microns. Simple!

Dave and I decided to measure ten size fractions for each sample. This required us to collect ten pipettes at exactly spaced time intervals. We had 40 sediment samples to analyze making a total of 400 pipettes. In order to measure the weight of sediment suspended in each pipette, a filter paper attached to a vacuum device was used. This meant we had to have 400 pre-weighed filter papers, with each paper carefully numbered and labeled with its weight entered onto a log sheet. The number crunching was all done by hand, of course, as this was in the age before personal computers.

Since we couldn't take 40 pipette samples simultaneously, we arranged them in pairs (one for each of us) with staggered starting times. However, the time between the first and second pipettes is only 25 minutes, which meant that we did not have quite enough time to start all 20 pairs before the second pipettes of the first batch had to be taken. Suffice it to say, things got a bit confusing!

The frenzied dance that ensued, as Dave and I frantically took pipettes of muddy water, ran them though the vacuum device, placed the used filter papers in a drying oven, and recorded the filter paper weight against time and sample number, is a memory I keep when I want to recall how we did things in "the good old days" or when I just want to laugh. As the experiment progressed, the time between pipettes grew. At one stage we started a series of 40 pipettes at midnight, and we did not finish until 3 AM! A week later the final batch of 40 pipettes was taken to determine the finest clay size fraction that still remained, slowly sinking through the murky water of those (now despised) graduated glass cylinders.

When we eventually finished reweighing the dried filter papers, we then needed to convert the ten weights for each sample into a grain size frequency distribution curve. Although this can be done manually, it is a laborious process, so Dave and I decided to put our computer skills to use, and we wrote out the code in Fortran, on punch cards, to run the calculations. Nowadays, the determination of sediment grain size is done using a laser device connected to a computer, and the grain size distribution in 32 size categories can be calculated in a few seconds. We could have run our 40 samples in an hour or so using a modern laser particle size analyzer, with replicates!

The end result of our experiment, you will be relieved to hear, is that the grain size does indeed become finer toward the center of the Dyes Inlet tidal gyre, as predicted by theory. We concluded that the mound is a product of the modern tidal current regime and not of glacial origin. In the grand industry that is human scientific endeavor, ours was but a small triumph. Even so, Dave and I were satisfied (and we got an A for our project!).

***

The receding glaciers left large valleys and fjords in Puget Sound. In other places, they left thick sediment deposits behind, like the ones in the North Sea of Western Europe. Here we find sandbanks that are many kilometers long and 10 to 20 m in height. Each large sandbank is surrounded by dunes, marching around them in a circle in regular ranks and files. They are composed almost entirely of glacial sand and gravel that has been reworked and sorted by strong tidal currents that flow over the seabed of the North Sea.

Living on and within these dunes are millions of small fishes called sand eels (sand lance). They dart in and out of the soft sand on the dune crests, eating small planktonic crustaceans (Copepods) that drift past in the current. From the safety of our hover car, we can watch as the sand lance, in turn, are hunted by salmon and cod. On the shallow dune crest, a puffin (or was it a cormorant, it moved too fast to see clearly) dives into the water and grabs a sand lance for breakfast. The current is getting stronger now, and we can see sand is drifting along the tops of the dunes and cascading down the steep slip face, just like we would see on a desert dune on land; the sand lance quickly bury themselves for shelter.

Sand eels are the favorite food for Atlantic cod. Once the backbone of the European fishing industry, cod stocks have been depleted to the point where fishing has been greatly reduced or banned in some locations. Strict management efforts have allowed North Sea cod to make a comeback in 2017, but scientists are urging a cautious approach to resumption of commercial fishing.

Looking back from our hover car, we see a large pipe hanging down from a ship floating over our heads, and it looks like a giant vacuum cleaner is sucking sand off the seabed! This is a sand dredger at work, a common sight along some coasts. A huge cloud drifts away from the ship as it sieves out the valuable coarse sand gravel and lets the fine mud wash overboard.

Sand is a valuable commodity!

If you were asked what natural resources you paid for today, you would probably list water and food, no doubt some oil-based products, and you could probably think of many other things to add to your list. But how much sand did you use today? The answer may come as a surprise because it turns out that, after water, sand is the second most used natural resource on our planet.[3]

We use sand mainly to make concrete for buildings, for road base, and for laying train tracks. Sand is mixed into asphalt and it is used as fill for construction. Globally, our annual sand and gravel consumption is estimated to be around 53 billion tons.

[3] Peduzzi (2014).

That's equal to around 7.5 tons/year (20 kg/day) for every person on Earth, enough to build a concrete wall 27 m (90 feet) tall and 27 m (90 feet) wide encircling the Earth (an analogy that may appeal to certain of our political leaders!). It turns out that the rivers in the world collectively supply around 16 billion tons of sand per year to the coast.[4] In other words, humans are currently using sand at a rate that is more than three times faster than it is being produced by nature.

Sand and gravel are mainly quarried on land, but as these resources run out or become too expensive to mine, people are turning to the sea to access sand. Desert sand is too round and smooth and no good for concrete. It is quite ironic that several Arabian countries import sand from as far away as Australia for their building enterprises. But many countries have turned to the ocean for their sand supply. Sea-based aggregate mining in the United Kingdom, for example, extracts around 20 million tons of sand and gravel each year from an area of seabed equal to about 1500 square kilometers (580 square miles). In Australia, the Brisbane airport was built using sand mined from sandbanks in Moreton Bay (Fig. 8.2).

Perhaps the greatest use of marine sand is for landfill and beach "nourishment." Global sea level has risen about 19 cm (7.5 inches) in the last 100 years due to human-induced climate change, and the rate of rise is increasing (it was rising about 1.7 mm/year at the end of the twentieth century and is now rising at around 3.4 mm/year), with the result that beaches are retreating landward. This is not acceptable to people living along the coast who own the property that would otherwise disappear underwater, so more and more sand is used each year to keep the beaches at a stable position above sea level. Billions of tons of sand have been pumped onto beaches along the US east coast over the last 25 years. In North Carolina, for example, the cost of beach nourishment over the last 25 years is estimated at over $700 million, and over $1.7 billion was spent on New Jersey's beaches over the same time period.[5]

The dollar cost of dredging is not the only price society pays. Dredging the seafloor is inherently destructive to marine life. It kills not only the animals that get caught by the dredge but the plume of disturbed mud settles over a broad area of the seafloor smothering many others. The disturbed areas may take decades to fully recover. And there is a trade-off here: sandbanks are where sand eels live, which is what cod eat, so removing sand for one industry must have an effect on the recovery of the cod stocks impacting on another industry.

***

Why are some beaches made of gravel and rocks while others are sandy or muddy tide flats or rocky and backed by cliffs? Marine geologists have spent a lot of time finding the answers to these questions, and we now know that the reasons why a particular coast is sandy or rocky depend on the interactions of many factors. The main ones are the presence of rivers delivering sediment to the coast, coastal elevation (steep and mountainous versus gentle and low-lying), and the energy of the adjacent ocean (wave regime, frequency, and intensity of storms and tidal range).

---

[4] Syvitski et al. (2005).

[5] http://beachnourishment.wcu.edu

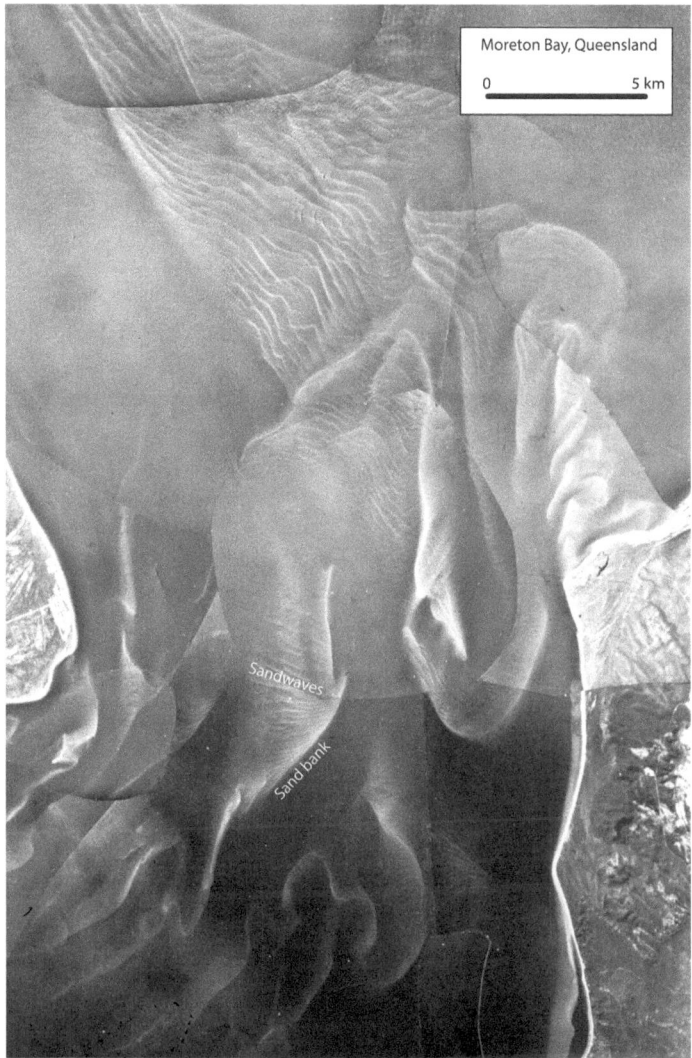

Fig. 8.2 Air photograph collage showing tidal sandbanks and sandwaves in the entrance to Moreton Bay, Queensland, Australia. About 11 million cubic meters of sand have been mined from these sandbanks in 2014 for sand used in construction for the Brisbane airport. (From Pattiaratchi and Harris 2002)

Consider what happens to a mixture of mud, sand, and gravel on the seabed during a storm (geologists are very particular in terms of particle size about the terms "mud," "sand," and "gravel" – "mud" is all particles less than 0.625 mm, sand is all particles between 0.625 and 2.0 mm, and "gravel" is everything larger than 2.0 mm in size). Waves and currents stir the sediment, first raising the smallest particles of mud into the water column. These smallest particles are carried along by the current, suspended in the water. As the storm intensifies, the waves get larger, and sediment grains of increasing sizes start to be moved along the seabed. Sand-sized grains are moved followed by larger gravel-sized grains. At the peak of the storm, only the largest gravel (cobbles and boulders), too big for the waves and currents to lift off the bottom, remains behind. This coarse, unmoved gravel layer is referred to as a "lag" deposit by geologists.

As the storm wanes, the wave and current power reduces, and the biggest, heaviest, gravel-sized grains stop moving first. These biggest grains may have only been rolled along the bottom rather than being lifted up into the water column. The movement of sand and gravel rolling and hopping along the bottom is called "bedload transport" by geologists. After the gravel stops moving, the sand-sized grains may continue moving for some time before they too eventually stop moving. Last of all, the tiniest particles that had been suspended in the water column (called "suspended sediment load" by geologists) gradually settle back to the seabed according to their size, from largest to smallest.

What happened to our original mixture of gravel sand and mud during the storm? The lag deposit made of the coarsest gravel was left behind, unmoved on the bottom, and stripped of all the smaller-sized grains. Gravel rolled and hopped away along the seabed for some distance. Sand rolled and hopped further still, and some fine sand may have been carried as suspended load even further along by storm currents. The finest mud particles may have remained suspended in the water column for many days after the storm, to be carried by currents for miles out into deep water. The storm has *sorted* the original sediment mixture into deposits of different size classes. Sorting of sediment by waves and currents provides the raw material to make beaches.

The process of sediment sorting happens whenever water (or air) flows over a substrate comprised of mixed particle sizes. It happens in rivers and deserts and in the ocean. Sorting is what makes our beaches sandy.

Storms occur along the coast several times a year, and each storm does its work sorting sediments into different sized deposits. The mud is removed and settles out in deep water, below the reach of storm waves and currents. Over the last 2 million years, sand from rivers that empty into the Tasman Sea along the southeastern coast of Australia has been sorted and transported north by storm waves and currents to create one of the largest areas of sand islands in the world in southern Queensland. River sediments are supplemented by erosion of sandstone cliffs occurring on the coast, and waves transport sand northward at a rate of 100,000 to 500,000 tons/year.

As sandy coastlines go, they don't get much sandier than the beaches and islands of southeastern Queensland. Over a stretch of coastline several hundred kilometers long, there is a string of huge islands made entirely of quartz sand. The largest, Fraser Island, is 120 km (75 miles) long and 24 km (15 miles) wide. Further south lies Moreton Island, which forms an embayment on its landward side where the Port of Brisbane is located. The highest sandhill in the world, Mount Tempest, 280 m (920 feet) in elevation, rises at the northern end of Moreton Island.

If you were to look down at the sea while flying into Brisbane airport, you could not help but notice the amazing complex of sandbanks located in the entrance to

Moreton Bay (Fig. 8.2). The tops of the sandbanks lie less than 2 m below the sea surface and can be seen plainly in air photographs and satellite images. The sandbank complex is estimated to contain 3.7 billion cubic meters of sand. Brisbane airport runways are built on 16 million cubic meters of sand mined from the sandbanks of Moreton Bay in 1983. Sand has also been dredged from Moreton Bay to build artificial islands used as port facilities. A further 11 million cubic meters was extracted in 2014 for airport extensions.

The extraction of sand has been locally controversial amid concerns over the impact on the environment. Removing sand from the sandbanks could cause local coastal erosion if it is not carefully managed. Nevertheless, the Queensland Government plans for the future involve mining another 40 million tons for expansion of the airport and Port of Brisbane facilities.[6]

Sand is indeed a valuable global commodity and the demand for sand is growing every year. Perhaps the biggest growth area is likely to be in mining offshore sandbanks to stabilize coasts and built infrastructure as the result of rising sea level. People will increasingly need to make decisions about the trade-offs between the demand for more sand versus the damage caused to seabed habitats from dredging operations.

***

Back in our hover car, we leave the dredged area of sandbanks behind. The view gets clearer, and we see a vast abundance of fish swimming over our heads, shoals of herring being chased by salmon. Sharks and even a whale swim by.

The gently undulating topography of sediment-covered seabed is occasionally interrupted by rocky outcrops and rough seabed strewn with cobbles and boulders. Overall, the continental shelf is mostly (over 90%) covered in sediment. Rocky habitats are rare. Where they do occur, rocks and boulders provide unique habitats for specialist species that have evolved to take advantage of their special qualities. Rocky surfaces are like islands among an ocean of sediment, producing oases of increased biodiversity.

Giant kelp forests are an example of ecosystems that only grow upon such rocky substrates. They attach themselves to stable rocky surfaces and grow upward toward the sunlight, making a canopy of floating, gas-filled bulbs and fronds. Kelp forests are the home to an array of species of fish and other animals that make use of the forest to shelter, feed, and reproduce.

Sadly, climate change has led to a dramatic decline in kelp forest around the world. Among places hardest hit is Tasmania (the island off the coast of southeast Australia) where warmer ocean temperatures have brought an invasive species of sea urchin. These invaders have consumed 95% of the kelp, leaving only barren rocks behind as they eat their way along the coast. From Alaska to California, the once extensive kelp forests along the west coast of the North America have been greatly reduced by grazing urchins whose populations were held in check by

---

[6] Queensland Government Department of Environment and Heritage Protection, Morten Bay Sand Extraction, https://www.ehp.qld.gov.au/coastal/regional-studies/moretonbay/

starfish. But the starfish, *Pisaster ochraceus*, the ochre star, is dying from a disease whose origin is unclear – scientists suspect climate change and warming of the northeast Pacific is to blame.[7]

Driving away from the kelp forest, we fly in our hover car once again over the broad area of shelf sediments. We travel carefully to avoid damaging the fragile sponges and anemones attached to the seabed, which seem to be growing only in isolated, elongate patches. The sea life within the sponge patches is rich and diverse, populated by different small fish, crustaceans, and other animals; it looks like a veritable "garden" of multicolored sponges, soft corals, and sea whips. In contrast, the seabed surrounding the patch of sponges is mobile sand ripples and sediments, mostly barren of life, with long furrows trending across the seabed.

Overhead a fishing boat cruises past towing a big net held down by two large steel plates (otter boards) that drag along the seabed, scooping up the sponges and anemones. Now we can see why the sponges and anemones are only found in a few elongate patches; they are the survivors! Otter boards dragged by the fishing boat dig long furrows through the sediments, while the chain or wire draped between the boards to hold the net down pulls up animals attached to the bottom. Dredges used to harvest scallops are even more destructive.

Studies show that between 6% and 41% of animals living on the seabed are trawled up with each pass of the fishing boat. Any fish or crustaceans larger than the mesh size of the net are also captured. All of the sponges, soft corals and sea whips, and some of the fish will be discarded – they are unwanted "bycatch." Most of the discarded animals do not survive and are thrown overboard to feed the seagulls or sink to the bottom.

Rocks and shipwrecks can snag fishing nets and ropes which the fishermen must cut loose and abandon. Lost crab pots and other fishing gear litter the seabed in the most popular fishing areas since about 10% of all fishing equipment is lost every year. The tragedy for fish and crustaceans is that the nets, hooks, and traps continue their work even when abandoned. This so-called ghost fishing must rank among humanity's most destructive and wasteful impacts on ocean wildlife. The amount of sea life killed in this way every year is unknown, but it is probably significant.

Fishing trawlers have destroyed the original habitat that was once a sponge garden across much of the Earth's shallow seas, especially along populated coastlines, leaving barren sediments behind. Much damage was done in the middle of the twentieth century when fishing effort expanded into frontier fishing areas. On Australia's remote North West Shelf, the seabed was prepared for trawling during the 1960s and 1970s by two fishing boats dragging a heavy chain between them over the bottom (pair trawls) to remove unwanted seabed life that would impede fishing. *Cutting the forest to catch the deer* seems like a good analogy.

After being trawled it can take from 2 to over 6 years for the trawled area of seabed to recover. But too often they never do. This is because bottom trawlers fish the same ground over and over, with the result that some pieces of seabed are trawled three or four times every year, which means trawl grounds are not allowed enough

---

[7] http://e360.yale.edu/features/as-oceans-warm-the-worlds-giant-kelp-forests-begin-to-disappear

time to ever fully recover. It is estimated that an area equal to the entire global continental shelf (around 32 million square kilometers or 12.4 million square miles) is trawled every year.[8] The global catch of wild fish is currently around 89 million tons per year, and 90% of it comes from the shallow continental shelf seas, and 20% of that (about 16 million tons per year) comes from bottom trawling.

Animals living on land and in the ocean are being decimated as humans eat them, hunt them for trophies, pollute their environment, and destroy their habitats. Global populations of fish, birds, mammals, amphibians, and reptiles were reduced by 58% between 1970 and 2012,[9] and their numbers are still in decline. Over this same time interval, the human population doubled, from 3.5 to seven billion. There is, in fact, a direct correlation between the increasing number of humans on Earth and the decreasing number of wild animals.

Pollution of the oceans includes the direct dumping of dredge spoil, sewage sludge, and industrial chemical waste. Added to this are excess or unwanted chemicals and debris that are dumped at sea or that wash in from land. Plastic is one of the most common forms of pollution that can be seen on the seabed. One remotely operated vehicle (ROV) survey on the mid-Atlantic ridge reported the "frequent occurrence of garbage (e.g., plastic bags and other objects) at all depths over very wide areas."[10] Seafloor observations from European waters between 50 and 2700 m water depth found maximum concentrations of garbage (mostly plastic bags and bottles) of over 100,000 pieces of debris per square kilometer.

Most plastic debris sinks to the seafloor and is eventually buried in the sediments. Plastic that floats is gradually broken down into small fragments. Tiny filaments of plastic are generated when clothes made of artificial fabric (like polar fleece) are in the washing machine – the microplastic particles end up in the ocean. Facial scrubs and other makeup products may contain microplastic abrasives that go down the sink and also end up in the ocean. About 8 million tons of plastic ends up in the ocean every year, and it is estimated that by the year 2050, the weight of plastic in the ocean will exceed the weight of fish!

The floating plastic accumulates in the mid-ocean gyres as dispersed small pieces and particles (but not as plastic "islands" as implied by some media reports) and images of plastic debris in the decayed bodies of seabirds and chicks have captured the public attention in recent years. I think it was the discovery of microplastics in seafood and sea salt products destined for human consumption that has finally prompted action. A resolution was passed in 2017 by the United Nations Environmental Assembly to curb marine debris and plastic pollution. It was probably inevitable that plastic would be found in humans; a study by doctors in Vienna has now discovered microplastic particles in human stools and concludes that over 50% of us probably have microplastic in our bodies.[11] Plastic pollution is no longer an hypothetical issue for marine animals; it is a matter concerning human health.

---

[8] Norse and Crowder (2005).

[9] http://wwf.panda.org/about_our_earth/all_publications/lpr_2016/

[10] Galgani et al. (2000).

[11] Schwable et al. (2018).

We can each of us take greater responsibility for taking care of the oceans and ultimately of ourselves. Each of us should be aware of the source of fish we eat and how we must curb our personal use of plastic. Another way to help species to survive the impacts of humans is to set aside protected areas in the sea where hunting and fishing are banned or strictly controlled. Parks on land are familiar to us, but parks in the ocean are a relatively new idea. Where marine parks (marine protected areas or MPAs) have been established, they have been shown to not only protect the environment but to also produce a fish surplus that overspills from the park, making adjacent fishing grounds more productive.

This is a key lesson for us all – conservation and industry must form partnerships (not competitions) in order for both to succeed. More efficient, productive, and prosperous fisheries can only exist where land-sourced pollution is curbed, where fishing practices are sustainable and include the protection of the fragile ecosystems upon which they depend. The United Nations has set a target of placing 10% of the oceans into MPAs by 2020. Achieving this goal will be a good start to reversing the decline in ocean biodiversity and fisheries.

Fishermen are often the first people to visit parts of the ocean while they search for new fishing grounds. An American seal hunter, Nathaniel Palmer, is thought to have been among the first people to sight the continent of Antarctica. The location of good fishing grounds is valuable commercial information, worth keeping secret, so fishermen are not likely to say too much about them in public. But such a secret, like the existence of previously unknown coral reefs, can sometimes leak out.

***

The bar of the Albatross Bay Resort in Weipa, in far north Queensland Australia, is a good place to catch up on what's been happening in the region and to hear the local gossip. Stories are told about long fishing trips into the Gulf of Carpentaria, and tips about the best fishing spots are traded. One story told by some halibut fishermen caught my attention back in 2003. It was a story about some rocky reef formations labeled by the fishermen as a "lost city" because of the way it appeared on their echo sounders, like tall vertical buildings standing upright on the seabed.

A lost city in the Gulf of Carpentaria may sound strange, but you need to first understand the background to this story. The Gulf is one of the flattest geographic regions on Earth. Between the towns of Weipa on the east side to Gove on the west side is a distance of some 500 km, and along that distance, the change in depth is a mere 65 m (213 feet). Out in the middle of the Gulf, the mean depth is around 50 m (150 feet), and there is not a single hill or valley more than a few meters in relief for hundreds of kilometers in any direction. Any small rock bump or dip in the seabed more than 5 or 10 m in relief really stands out.

Nautical charts show a number of small shoals scattered across the southern part of the Gulf. They don't look like they would be anything special from what can be seen on the chart, but when you cross over them with an echo sounder, their near-vertical sides stand out like the walls of buildings. The contrast with the flat expanse of the Gulf could not be greater. From a depth of 50 m (150 feet), these rocky "buildings" rise over 20 m (60 feet) to a depth of around 28 m (91 feet). Their tops are mostly flat apart from some rugged bumps along the edges. What could these features be?

Our survey to the Gulf of Carpentaria in May 2003 aboard the RV *Southern Surveyor* was intended to map the seabed sediments in the Gulf and determine their origin. I was working at that time with a team from Geoscience Australia, the national geoscience agency, which has its headquarters in Canberra. The scientific party joined the ship in Weipa (which is how I happened to overhear the "lost city" story at the Albatross bar). As an aside, we decided to carry out some exploratory work over one of the curious shoals marked on the chart.

On board, we had a multibeam sonar mapping system (as discussed in Chap. 4) which in 50 m water depth allowed a 250 m (750 feet) wide swath of seabed to be mapped with soundings spaced 2.5 m (8 feet) apart. After about 2 days, we had mapped an area of around 100 square kilometers (39 square miles) to create a precise, three-dimensional image that revealed the true character of these mysterious shoals.

And we discovered that they are submerged coral reefs!

Lowering a camera onto the reef top, we found patches of live corals growing. The discovery of coral reefs living in the Gulf of Carpentaria was a sensation and made the news in Australia. It seemed bizarre to people that huge coral reefs were still undiscovered. The reason is simple enough – they are submerged 28 m (91 feet) below the surface and could not be seen from above.

The Great Barrier Reef grows along a 2000 km stretch of the northeastern margin of Australia and can be seen from space, but reefs do not grow near the sea surface in the Gulf of Carpentaria. The Gulf of Carpentaria coral reefs went unnoticed because they cannot be seen in satellite images or from an airplane. They grow exclusively at a depth of around 28 m (91 feet), hidden beneath the sea surface. These are "mesophotic" coral reefs (meaning "middle light" because of the lower levels of sunlight that coincides with the depth where they are found). If an entire reef province can exist hidden in plain sight in the Gulf of Carpentaria, how many others might there be in the world ocean? The answer is: probably quite a lot. The true distribution of mesophotic coral reefs remains hidden from us, and mostly they will remain among the oceans many mysteries.

During a second survey to the Gulf in 2005, we sampled the reefs using a rotary rock drill. This machine comprised a four-legged, weighted platform with a 3 m (10 feet) tall derrick on top. The drill had a 75 mm (3 inch) diameter pipe and was driven by a hydraulic pump which ran from an electric cable tethered to the ship. The crew shook their heads in disbelief when they saw what looked like a lunar lander from the Apollo missions being loaded aboard the ship (Fig. 8.3). Their experience with such complex-looking equipment designed by scientists is that it usually ended in dismal failure.

I am happy to say on this occasion the crew's skepticism was disproven. The drill worked very well, and the drill cores proved that the reefs are comprised of coral limestone. The whole of these mound-shaped features, 25 m (75 feet) tall and covering over 100 square kilometers (39 square miles), are made of dead corals, the

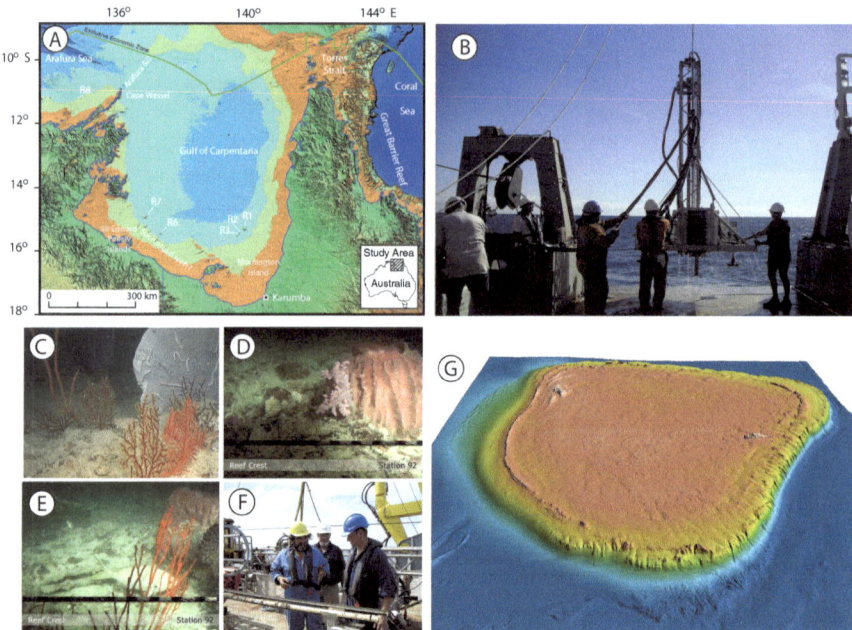

Fig. 8.3 Gulf of Carpentaria survey campaign, 2003–2005. (**a**) Location map showing submerged coral reefs R1 to R8; (**b**) rock drill deployed from back deck of RV *Southern Surveyor* in 2005; (**c–e**) underwater photographs of coral, sponge, and fish; (**f**) Andrew Heap (right) and Jon Stratton inspecting drill core of reef limestone; and (**g**) three-dimensional multibeam sonar map of the largest submerged reefs R1, showing its raised rim and upper surface at a depth of 28 m

same as most other reefs. Radiometric dating of the corals indicated that the reefs had grown in a brief spurt just as rising sea level flooded the Gulf around 10,000 years ago but had since slowed down. In all we mapped 7 of the largest shoals and collected nearly 50 drill cores and many hours of underwater video to document a new coral reef province, previously unknown in Australia.[12] Later we showed that around 50% of the Great Barrier Reef is also growing at mesophotic depths – only about half of the Great Barrier Reef can be seen in satellite images. The rest lies hidden beneath the waves.

Based on our work, the Australian government subsequently placed the submerged Carpentaria coral reefs within marine protected areas in order to conserve them. The Carpentaria marine reserve network will protect 23,775 square kilometers (9180 square miles) of seabed habitats, including submerged coral reefs.[13]

Sadly, in the last decade, global warming has killed many coral reefs. Overheating of the surface water causes the coral polyps to eject their symbiotic algae, which turns the coral skeleton white, described as "bleaching." The worst coral bleaching on record has taken place in 3 years, 2015–2017.

---

[12] Harris et al. (2008).

[13] http://www.environment.gov.au/topics/marine/marine-reserves/north/gulf-of-carpentaria

The United States National Oceanic and Atmospheric Administration (NOAA) declared 2015–2016 to be the third global coral bleaching event, following similar events in 1998 and 2010. In 2016 and again in 2017, approximately 50% of Australia's Great Barrier Reef was bleached, most severely in the northern part, where two-thirds of the corals that were bleached have died. Globally, bleaching was worse in some locales (e.g., Kiribati) and affected some areas that had not bleached in the previous events (e.g., the northern Great Barrier Reef).

Coral bleaching has become a regular event, occurring on average once every 6 years. But it takes over 10 years on average for a reef to recover from a severe bleaching event. Anthropogenic climate change has exceeded a critical tipping point for coral reefs because human activity has now warmed the oceans to a temperature whereby combinations of El Niño with other factors will cause widespread coral bleaching most years from now on someplace in the ocean. The concept of tipping points for natural systems is well established in the scientific literature. An "ecological tipping point" has been defined by the Biodiversity Information System for Europe as a "situation in which an ecosystem experiences a shift to a new state, with significant changes to biodiversity and the services to people it underpins, at a regional or global scale." Passing a tipping point means that a "threshold beyond which an abrupt shift of ecological state" has occurred, causing long-lasting changes that are hard to reverse.

The bleaching and death of vast areas of coral reefs in recent years – in the northern and southern hemispheres – seem to fulfil this definition. Dead coral reefs can recover if allowed enough time and depending on other environmental factors. But when the frequency of bleaching events exceeds the rate of recovery, a tipping point has passed.

Why has the tipping point been passed now? What is so special about the present ocean temperature and atmospheric CO_2 levels? One of the factors that makes coral bleaching hard to understand and predict is related to the lag time between a rise in carbon dioxide in the global atmosphere and the warming of the oceans. This lag effect has huge implications for coral reefs and the future of our oceans.

***

Put a pan of cold water on the stove, and turn the heat up all the way and what happens? Nothing at first. But wait for 5 minutes and the water will boil.

The time difference between turning up the heat on the stove and the water coming to boil is explained by the *heat capacity* of water, the number of calories needed to raise the water from room temperature to boiling point. The same effect occurs in the ocean. Global climate change is causing the atmosphere to warm up, and the extra heat is transferred to the ocean. But the heat capacity of the oceans is enormous.

The top 3 m of the ocean has the same heat capacity as the entire atmosphere. And the oceans are 4 km deep on average. This means that the oceans can store a lot of heat. And because of that, there is a lag time between warming of the atmosphere (which heats up practically instantaneously) and the warming of the ocean.

We know that rising levels of carbon dioxide in the atmosphere are the main cause of global warming. As the carbon dioxide concentration increases, the greenhouse effect causes more longwave radiation to be trapped in the atmosphere, which becomes warmer. This heat is transferred first to the surface of the ocean and eventually into its deeper reaches, but it takes a long time. Decades, in fact.

Ocean warming lags behind global atmospheric carbon dioxide (CO_2) levels. It takes around 25–50 years for the ocean to absorb about 60% of the excess heat generated by global warming,[14] and it will take perhaps a century for the ocean to absorb 90% of the excess heat. This means that, as CO_2 levels rise, the atmosphere warms quickly, but ocean warming occurs gradually with the full effect being realized only many decades later.

To put this in context, Charlie Veron and colleagues at the Australian Institute of Marine Science predicted that a coral bleaching tipping point occurred when atmospheric carbon dioxide levels exceeded around 350 parts per million (ppm).[15] But we passed 350 ppm 30 years ago, back in 1988, and we are only now in the midst of a coral bleaching crisis that has affected 70% of the world's reefs in the last 3 years. That's because it has taken the oceans 30 years to warm up in response to those long-past atmospheric CO_2 levels.

Meanwhile, CO_2 levels have now passed 400 ppm and the warming continues. Even if we could halt CO_2 levels at their current levels, it would take 25–50 years before just 60% of the full warming effect will take place. There is nothing we can do to stop this ocean warming from occurring. The future ocean will be much warmer than today's temperatures, and these have already passed the tipping point for tropical coral reefs.

That's the lag effect. It is one of the most insidious aspects of climate change. Inaction on climate change in the late 1980s effectively condemned coral reefs to death by bleaching 30 years later. We will not have to pay the full price for our inaction on climate change today – that cost will be borne by our children and our grandchildren, once the ocean begins to equilibrate with today's more than 400 ppm CO_2 level.

Another effect of the CO_2 dissolved into the ocean is that it makes seawater more acidic, which adds yet another burden to corals struggling to survive. We know now that the biological effects of acidification are far-reaching and are of growing concern to marine scientists (we'll come back to this topic in the next chapter).

Many of the reefs we have now will not exist in the future – many will not survive, but some coral species will be able to adapt to warmer, more acidic oceans. Corals are unlikely to go extinct because as the oceans warm, corals will be able to colonize newly available habitat at higher latitudes. It will take time, but eventually new reefs will grow in locations that were once too cold. Their transition will need to be carefully managed if reef ecosystems are to survive to the end of this century.[16]

---

[14] Hansen et al. (2004).

[15] Veron et al. (2009).

[16] Hughes et al. (2017).

***

Globally, coral reefs cover an area of around 250,000 square kilometers (97,000 square miles) and, according to the United Nations World Ocean Assessment, are of major importance for 275 million people located in 80 countries who depend on reef-associated fisheries as their major source of animal protein. Coral reefs are the rare and precious jewels of the ocean, found only on about 1% of continental shelves. Their biodiversity rivals that of tropical rainforests and the benefits they provide to people have been collectively valued at USD $29 billion per year. This includes their value to tourism ($11.5 billion), fisheries ($6.5 billion), and coastal protection ($10.7 billion). The demise of reefs puts all this at risk and threatens the health and prosperity of people who depend on them.

Once the corals have died, they no longer grow vertically upward. The reefs gradually will erode and become less effective in providing shoreline protection from wave action during storms. Dead corals not only lack the aesthetic appeal that is fundamental to reef tourism; they also do not sustain the same fish community. This results in reduced tourist activity and reduced income from fisheries that can threaten the livelihoods of local communities. Living coral reefs are important religious symbols for some communities, and the loss of corals harms this important cultural aspect as well.

If coral reefs really are the "canary in the global climate change coal mine," then it seems the canary may have just dropped off its perch! Talks to reduce greenhouse gas emissions by 2050 will be too little too late as far as coral reefs are concerned.

What can be done is to try and protect the reefs most likely to survive anthropogenic climate change. One idea is to identify the coral reefs that might survive and encourage governments to set these reefs aside for protection and conservation.

But which reefs?

One reef type worth considering could be mesophotic reefs, like the ones discovered in the Gulf of Carpentaria. The reason is because these reefs are protected by the layer of water lying above them. The overheated water is often confined to the uppermost surface layers, with cooler waters located below. In this way, the mesophotic corals might survive while their shallow water cousins succumb to the heat. If and when conditions are right, polyps from the deep reefs might recolonize the shallow sites allowing the reefs to recover. This is the idea behind the "coral reef lifeboats" concept.[17]

There is a long way to go to protect the world's coral reefs. Although about 27% of coral reefs are located in marine protected areas (MPAs), most of the MPAs are not adequately enforced or do not offer full protection to the corals, such that it is estimated that only around 6% of reefs are actually protected in functioning MPAs.[18] It is a race against time to locate and protect coral reef lifeboats.

***

---

[17] Baker et al. (2016).
[18] Burke et al. (2011).

The power of the oceans is impressive. Anyone who has witnessed huge ocean rollers crashing onto the coast after a storm cannot but stand in awe. If only we could somehow harness that power and turn it into energy for human use, all of our electricity needs could be met.

In fact, estimates of total global wave energy and current global use of electricity are both around 2 terawatts. In other words, wave energy alone could theoretically provide all the power we need.

The idea of extracting renewable energy from the oceans is not new, and there are several ways that the ocean's power can be harnessed. Waves crashing on the coast not only appear impressive; they contain useful energy which, in recent times, people have built machines to extract. However, the first renewable energy machines were built along coasts that have large tidal ranges where swift tidal currents occur in narrow passages such as occur between islands and the entrances to bays or estuaries. The tides along the coasts of Western Europe attain over 6 m (20 feet) in height in many areas, and it is not surprising that the idea of building a dam and water wheel to run a mill next to a tidal race occurred to people living there long ago. In Suffolk, England, there is a tide mill that dates back to 1170.

On the southwest coast of Great Britain, the Bristol Channel funnels the ocean tidal wave, building it in height as it moves from the outer Channel and into the Severn Estuary, where spring tides reach to over 14 m (46 feet) in height at the town of Bristol. This tidal range is the second only to the Bay of Fundy in Canada, which has a maximum tidal range of 16.3 m (53 feet). The currents in the Bristol Channel are incredibly strong, over five knots (nearly 10 km/hour) in places, and there are several locations where engineers have proposed that would be suitable for building a tidal dam or barrage, to harness the tidal flow to generate electricity. Tidal power from the Bristol Channel could supply up to 5% of the UK's energy needs.

An ideal spot for a tidal barrage has been proposed for the region around Flat Holm and Steep Holm, two rocky islets that form a natural barrier between the inner and outer Bristol Channel. Over the years there have been other locations suggested, but most of the proposed designs involve building a dam or barrage of some kind. Since all rivers carry sediment, the reservoir behind a dam will eventually fill up with mud and sand, which can affect the performance of the dam and add to maintenance costs. On a global scale, the sediment trapped behind dams (3.7 billion tons) actually exceeds the amount of sediment released due to human-induced soil erosion (2.3 billion tons). Dam building has consequently resulted in coastal erosion in many places around the world.

The same sediment-trapping problem applies to tidal power dams, since the natural tidal current system transports marine sand into the estuary and disperses the sediment delivered to the coast by rivers. Before any tidal barrage could be built in the Bristol Channel, these questions had to be answered: What is the quantity of sediment naturally in motion within the Channel? What would happen to this sediment if a dam were built? Would the dam cause local erosion or deposition of sediment that could harm the environment?

To find answers to these questions became the aim of a PhD project that I started in the fall of 1981 at the University of Wales in Swansea, United Kingdom. Professor

Mike Collins, a friend and colleague of Prof. Dick Sternberg from the University of Washington in Seattle, had agreed to supervise the project, and he arranged for me to have access to some of the data collected by the engineering companies involved in the environmental studies of the proposed tidal barrage. This data set included over 3000 km (1860 miles) of side scan sonar data, used to map the rocks and sediment on the seafloor.

Side scan sonar is a technology that is based on measuring the reflection of sound waves that bounce off of the seabed. Anyone who has used an echo sounder on a yacht or fishing boat will be familiar with how sonar works: sound waves are reflected from the seabed to tell the water depth. Side scan sonar uses the same principle, but with two sonar transducers tipped on their sides and mounted in a tow fish (a torpedo-shaped body that is towed on a cable behind a boat). As the survey vessel steams along, sonar beams scan out to either side of the tow fish, pinging typically four times per second, to build an image of the seabed. Hard and rough rocky surfaces give a strong echo, making a dark tone on the sonar record, whereas mud and smooth bottoms give weak reflections and make light toned sonar records. The shapes of objects like shipwrecks, underwater sand dunes, or rocky reefs can be detected and mapped using side scan sonar.

Sonar data revealed that the Bristol Channel seafloor has a mostly rough and rocky bottom, with a single large submarine valley running down the middle. Sandy patches are found along the sides of the channel, filling the fringing bays with large tidal sandbanks (Fig. 8.4). The deep valley floor running roughly down the middle of the Bristol Channel is also sandy. Everywhere else was exposed bedrock, boulders, and gravel.

The next step was to work out which way and how fast the sand is moving within and along the sides of the Channel. To do this Mike arranged for us to borrow 12 current meters from the British Institute of Oceanographic Sciences (IOS), along with IOS technicians to help install and recover them. Aanderaa current meters are standard oceanographic equipment and used right up to the late 1990s when they were supplanted by more modern electromagnetic devices. The main component of an Aanderaa current meter is the Savonius rotor on top; a magnet on the rotor trips an electromagnetic counter located inside the pressure housing. The number of rotations per minute is counted and transformed into an average current speed.

Twelve current meters were placed in pairs at six locations on the Bristol Channel seafloor. Each pair was anchored to the bottom using an old railway wheel and suspended beneath a glass sphere encased in protective plastic covers. The mooring location was marked by a 2 m (6 feet) tall, cigar-shaped, fiberglass buoy, painted yellow, and equipped with an aluminum mast, flashing light, and radar reflector.

To make sure that everyone knew about the experiment, I mailed out letters to all the local authorities and commercial fishing operators. The locations of the buoys were advertised in maritime industry offices for 1 month before they were deployed, the coastguard was notified, and flyers were mailed out to local fishermen.

In early August of 1983, I was filled with scientific zeal and enthusiasm as I arrived at the dock where Swansea University's research vessel, the RV *Venturous*, was moored. I explained my plans to the *Venturous'* captain, Brian Lewis, and the

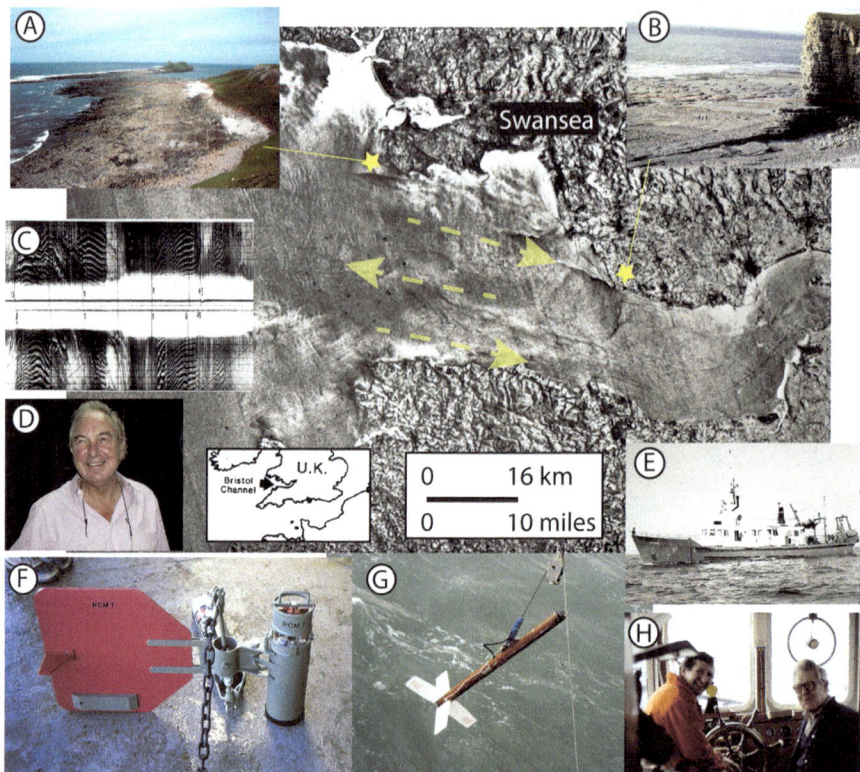

Fig. 8.4 Bristol Channel project 1981–1983. Main map is a synthetic aperture radar image of the Bristol Channel with location of Swansea. Large sand dunes and sandbanks are visible in Swansea Bay and off headlands like Worms Head and Nash Point. Arrows indicate mutually evasive ebb-dominated sediment exiting the channel and flood-dominated sediment transport into the channel in zones close to the coast. (**a**) photograph of Worms Head; (**b**) photograph of Nash Point; (**c**) example of a side scan sonar seafloor image showing sand dunes and rock in the central Channel south of Nash Point; (**d**) photograph of the late Prof. Michael Collins, photo from 2010; (**e**) RV *Venturous*, 35 m research vessel operated by Swansea University in the 1980s; (**f**) Aanderaa current meter; (**g**) side scan sonar tow fish; (**h**) Capt. Brian Lewis (left) and crew on the bridge of RV *Venturous*

crew, who gave me skeptical looks, shaking their heads at yet another impossible scientific scheme they must have heard from many students before. My plan was to leave the current meters out for one full month to record current speed and direction in the busy shipping lanes of the Bristol Channel. The meters were to be deployed with the help of Phil Taylor from the IOS technical support team based in Cardiff's seaport, Barry. At each station, the ship would heave to over the chosen site. The meters were to be switched on and the details recorded on log sheets. Then the weights were to be lifted over the side of the ship and lowered to the bottom.

The deployment went smoothly in beautiful, calm weather that can sometimes occur in August in the United Kingdom. I slept soundly every night through that

August of 1983 knowing that my current meters were out there under the Bristol Channel, ticking away, recording a current measurement every 5 minutes on reel-to-reel magnetic tape, and winding gently onto the plastic spools inside each meter.

The sunshine was bright, and the sky was blue when we set out 4 weeks later in early September to recover my current meters. The first meter was placed right outside Swansea Bay, just an hour steaming from the harbor. There was the yellow cigar buoy, but I noticed the radar reflector was missing. And there was a dent on the side of the buoy, where a boat or ship had accidently struck it. Obviously, there was one foolish sea captain who had not read my notice to mariners about the moorings! Ah well, not to worry. We pulled up the two current meters on the mooring, and there was no other damage – all was well.

Or so it seemed.

The next cigar buoy had not lost its mast, but the mast was bent over, the light was smashed, and the buoy was low in the water with a puncture in its side and half filled with seawater. The third buoy was also damaged, but again the instruments were pulled safely aboard the *Venturous*. Of the next three buoys, there was no sign. They were gone! Apparently run down by careless sailors who had taken no notice of my advertisements. My project seemed ruined and I was devastated!

The *Venturous'* captain, Brian Lewis, told me not to worry that we would wait till slack water and the cigar buoys, even if ruptured, may still rise to the surface when the tidal currents slowed. The *Venturous* steamed slowly ahead into the current, and we waited for slack water, while the crew gave me "I told you so" looks, certain that these current meters were doomed and never to be seen again. But luck was with us that day. As the current slowed near low tide, the ghostly shape of one cigar float rose to the surface, and we were able to tie on a cable to haul it aboard. The fourth mooring was safely aboard.

The 2-week spring-neap tidal cycle meant that our chances of finding the last two current meter moorings would be greatest at slack water of a neap tide. We waited for 4 more days and returned to the fifth site, located off Nash Point. Slack water arrived, and we steamed slowly over the site. We had brought with us the University's own side scan sonar, as our electronics technician, Mike Punter, suggested we might be able to locate the instrument using sonar (Fig. 8.4). There was no sign of the buoy at low tide, and we could not see anything in the sonar data either.

With darkness falling, on the final pass over the site with our side scan fish, Captain Brian said to me "sorry lad, I am afraid that one is lost." We stopped to pull in the tow fish, and I glanced over the side, and there, looming out of the depths, was the yellow cigar buoy, just below the sea surface! Thinking quickly the mate, Dai Harris, tied a large orange float to the wreckage of the broken radar mast just as the tidal flow started again, and the cigar buoy disappeared into the depths, dragging the orange float down with it.

We returned a day later at slack water and spotted the orange float struggling toward the surface. After a few hours of hard work using the ship's winch, we recovered the damaged, water-filled floats from the mooring. Most importantly, we also retrieved the vital data recorded by the two current meters.

It was a hard lesson to learn about collecting data in a busy shipping area like the Bristol Channel; all six of my moorings had been struck by one or more vessels, three had been sunk, and one was lost entirely. If I had tried for a longer deployment period, of say 2 months, I might have lost all of the moorings. On the positive side, I was able to recover 10 out of the 12 current meters, Phil Taylor and the IOS were happy with that, and I could get on with analyzing the data.

The current meter data showed that combined waves and tides move about 6 million tons of sand in and out of the Bristol Channel every year.[19] That's an amount of sand equal to about 60 supertanker loads. It is also a lot of sand compared with, for example, the input to the Bristol Channel from rivers, which Mike Collins estimated is about 1.6 million tons per year. The potential environmental consequences of interrupting this flux of sand include local coastal erosion and loss of coastal habitats. In the end, the British Government decided not to go ahead with building a tidal barrage across the Bristol Channel. I don't know if my current measurements were a part of taking that decision, but I do hope that, one day, someone will be able to build a tide power plant to harness the inexhaustible tidal energy of the Bristol Channel.

Renewable energy, from solar panels, wind farms, wave, and tide power plants, must eventually replace fossil fuels in order to reduce humanity's greenhouse gas emissions. This kind of transition has occurred before. In the United States, 1883 was the year when coal consumption exceeded wood as an energy source, and 1949 was the year when oil consumption exceeded coal as an energy source. The year must come when renewable energy will exceed both coal and oil as humanity's most important energy source.

There is an epitaph to this story: about 4 months after we had completed pulling up the five Bristol Channel current meter moorings, there was a knock on the door of Phil Taylor's office in Barry. A fisherman had "found" one of our current meters, claiming it had washed up on a beach and he had just seen it while out walking. He offered to sell it back to the IOS for a salvage fee. Phil said he'd have to take a look inside first to see if the meter was worth anything. Phil took the bashed and broken meter into a back room, opened the pressure housing, and retrieved the reel-to-reel magnetic tape containing the vital data (I did after all get the data from 11 out of original 12 current meters!). He then returned the wrecked meter to the fisherman and said he was sorry, but the meter was too badly damaged to ever be restored, so he could keep it as a souvenir!

***

Of the 32 largest cities in the world, 22 are located in estuaries. About 60% of the world's population live alongside estuaries, on river deltas, and other coastal settings. Estuaries and deltas are the outlets of the world's rivers and also the end point for most sewer pipes. Everything we put into rivers comes out at the coast into an estuary or delta. Even things we don't deliberately dump into rivers but put on the

---

[19] Harris and Collins (1988).

ground, from pesticides to plastic bags, inevitably get washed into a river and make their way to the coast.

The list of our transgressions against estuaries is long and depressing. The strange thing is that people choose to live on the coast over other locations. Many of us dream of owning a house overlooking a harbor. It is a paradox that estuaries are at once the most valued and cherished real estate and the most polluted and heavily impacted parts of the ocean.

Estuaries are a rare kind of habitat because they only exist briefly in geologic time, until they have been filled in with sediment supplied by rivers and washed in from the sea. Estuaries are ephemeral features, transitory in space and time. The story of an estuary begins when sea level is lowered, like it was during the last ice age 20,000 years ago. A river erodes a valley and carries the sediment down to the ocean. When sea level rises, the valley is flooded, and we have an estuary (at high latitudes, the valley is carved by a glacier, and it becomes a fjord when flooded by rising sea level).

If the river that feeds into an estuary carries a large sediment load, then the estuary may quickly fill in and begin to prograde into the ocean making a river delta. Human history is very short compared with the cycles of the ice ages, and we happen to live at a time when there are still many estuaries in existence. But it is the geological destiny of all estuaries to eventually become filled with sediment and to become deltas. Every delta has an estuary buried beneath. It is only a question of time.

The stories of river deltas fill our literature, and some deltas have been studied by historians and anthropologists as well as by marine geologists: examples include the Nile, Mississippi, Ganges, Volga, Huang He, and Danube, among others. In a benchmark paper published in 1975, William Galloway proposed that there are three types of delta, each dominated by one of three fundamental governing processes: rivers, waves, and tides.[20] We'll need to hop into our hover car and fly up high to get a bird's-eye view of these deltas.

From a great height, river-dominated deltas have a particular shape that looks like a bird's foot. The river looks like it is building long narrow "claws" where it is depositing sediment via multiple distributary channels. The bird's foot shape is how the Mississippi River delta looks from space.

Wave-dominated deltas are smeared out along the coast by the effect of waves eroding and transporting sediment, sometimes building barrier islands or sandbars offshore, aligned parallel to the coast. The coast is sandy, because waves have washed the finer muddy sediments offshore. The Nile delta is a good example of a wave-dominated delta (Galloway suggested the Copper River in Alaska was the best example). Both the river- and wave-dominated deltas have been studied in great detail, but tide-dominated deltas have received little attention. Galloway proposed the best example of a tide-dominated delta is Fly River Delta in southern Papua New Guinea, but up until the 1990s, there had not been a lot of research carried out on the Fly River.

One reason why is simply that the Fly River Delta is very remote. The nearest town in Papua New Guinea is Port Moresby located over 400 km (249 miles) to the

---

[20] Galloway (1975).

east. The nearest town in Australia is Thursday Island in Torres Strait, located 200 km (124 miles) to the south.

Another reason for the lack of exploration is that the Fly Delta is uncharted for navigation. The mud banks and islands of the Fly Delta are constantly changing in response to the river input of 85 million tons of sediment per year, and so any charts more than a few years old are likely to be wrong. The turbid waters of the Fly Delta are also extremely shallow – across the 100 km (63 miles) width of the three main distributary channels, the water depth averages about 5 m (15 feet) and is never more than about 10 m (30 feet). This is a dangerous place to take a boat!

The biggest environmental threat facing the Fly Delta is the Ok Tedi gold and copper mine that operates in the catchment of the Fly River. High annual rainfall averaging around 10,000 mm (390 inches) per year makes the steep mountain slopes unstable in the vicinity of the mine. The mine's tailings dam collapsed in 1984 within months of being completed, but the Papua New Guinea government allowed the mine to continue to operate by disposing of 80 million tons per year of mine waste directly into the river. The resulting pollution caused an environmental disaster, contaminating over 1000 km (620 miles) length of the river and killing an area of 1600 square kilometers (600 square miles) of tropical forest. A question we asked ourselves was: "Is there any evidence that heavy metals have escaped from the Delta to pollute the northern Great Barrier Reef?"

Four expeditions to the Fly Delta were made by my research group at the University of Sydney between 1990 and 1994. Our expeditions required the transport of all our equipment 3000 km (1860 miles) from Sydney up to Thursday Island in Torres Strait, where the boat picked us up. Evenings spent at the Federal Hotel gave us city slickers from Sydney a taste of the local culture, listening to the Mills Sisters singing *My Island Home*.

Our first expedition to the Fly Delta was aboard the HMAS *Cook*, the Royal Australian Navy's oceanographic research vessel, during her final cruise in April, 1990 (she was decommissioned soon after our survey was completed). The *Cook* was too large a vessel to get into the Delta's distributary channels, so our work was restricted to the deeper, seaward parts. The first seismic data were collected across the front of the Delta, showing it to contain "clinoforms," a kind of bed that is prograding seaward. Sediment core samples contained seasonal layers (varves) and sedimentation rates measured using radioisotopes showed that the delta front was advancing seaward at a rate of about 6 m (20 feet) per year.[21]

Our team returned the next year in 1991, this time with a shallow-draft, sailing catamaran, the RV *Sunbird*, to explore the shallow parts of the Delta's distributary channels and some distance up the main river channel. We documented tidal sediments on intertidal flats of the mud islands, upon which dense mangrove forests are growing. The islands are constantly changing, eroding away on one side while accreting mud on the other. Tall, straight-trunked mangroves grow up to 50 m (150 feet) tall; their trunks fall into the muddy waters of the distributary channels

---

[21] Harris et al. (1993).

Fig. 8.5 Fly River Delta project 1990–1994, arrows on the location map show the Coral Sea Coastal Current deflecting sediment exported from the Fly River toward the north, away from the Great Barrier Reef. Photographs show (**a**) mangrove forested deltaic coast with clear water against the shore and turbid water mid-channel; (**b**) HMAS Cook; (**c**) intertidal mudflats of the delta; (**d**) boomer seismic sound source in a raft for towing behind a ship; (**e**) seismic profiles of the rapidly prograding delta front; (**f**) Bob Dalrymple demonstrates to Jane Golding how to operate a vibrocorer; (**g**) eroding deltaic island with loose logs and falling mangroves; (**h**) scientific party of our 1993 survey, standing from left Jim Gardner, Peter Harris, Alison Cole, David Mitchell, Elaine Baker, Chari Pattiaratchi (in shadow), Phil Gibbs, Bob Dalrymple, Will Schroeder, and Jock Keene (kneeling)

from the eroding side of the deltaic islands (Fig. 8.5). Sailors have reported seeing tree trunks floating miles out to sea in the Gulf of Papua.

The *Cook* and *Sunbird* expeditions raised several questions about the fate of sediments discharged from the Fly Delta. Based on the sampling we had completed up that point, only about 47 million tons per year of sediment could be accounted for in our sediment budget for the Delta. Since the total load of sediment was estimated to be about 85 million tons per year, what happened to the other 40 million tons?

Our third expedition was to the Gulf of Papua located directly offshore from the Fly Delta, with the aim of tracking the missing sediment and assessing the potential damage caused by any pollution. The Gulf of Papua marks the northern limit of the Great Barrier Reef. Here is the transition from clear water and coral reefs to the muddy deltaic coastline of southern Papua New Guinea. Our expedition in April, 1993, used the Australian Research Vessel, RV *Franklin*, and we mapped and sampled sediments from the Delta front seaward to the continental shelf edge (Fig. 8.5).

Our work showed that there is very little sediment today getting offshore from the Fly into the northern Great Barrier Reef. This is because there is a strong, northward-flowing current, the Coral Sea Coastal Current, that sweeps any sediments (and any heavy metals) away from the corals and deposits it along the Papua New Guinea coastline. The Great Barrier Reef is protected by the Coral Sea Coastal Current from being blanketed and buried by the muddy Papua New Guinea rivers that discharge their massive sediment loads just to the north. Analysis of sediment samples by Elaine Baker and Alison Cole showed there was no excess of copper, cadmium, or other heavy metals detected in the Torres Strait or Great Barrier Reef sediments.[22]

The effectiveness of the Coral Sea Coastal Current in keeping the waters in the Gulf of Papua clear and mud-free was revealed by a calcareous alga called *Halimeda*. This remarkable algal plant produces segments made of calcium carbonate rendering it inedible to most herbivores. When dead, the segments, which look like cornflakes, pile up into thick mounds that can cover broad areas of the continental shelf. The Coral Sea Coastal Current keeps waters in the Gulf of Papua so clear that *Halimeda* (which requires sunlight) can grow to depths of at least 100 m (300 feet). The discovery of *Halimeda* growing in such deep water in such an unlikely place (in front of a muddy river delta) inspired our science team to sing the "*Halimeda* chorus" (sung to the tune of the Hallelujah chorus) whenever it was retrieved in our grab samples. Our lusty, out-of-tune singing confirmed our status as "boffins" (equal to scientific nerd) in the minds of *Franklin's* crew.

For our final survey, Elaine negotiated a deal with Ok Tedi Mining Ltd. to let us use their vessel, the M.V. *Western Venturer*, in January 1994. The *Western Venturer* was a small catamaran hull motor boat that was ideal for working in the Fly Delta. Since the bottom was constantly changing, the *Western Venturer* regularly ran aground, which didn't seem to bother the captain or crew (in fact I wondered sometimes if the crew ran the boat aground on purpose so they could have a rest until the tide came in again!). On this survey, we collected more samples and seismic data,

---

[22] Baker and Harris (1991).

and we made measurements of the currents and turbidity (amount of suspended sediment) of the water.

There are three main distributary channels of the Fly Delta, and we wanted to find out which of the three channels carried the most sediment to the sea. Strong tidal currents within the Delta keep so much sediment in suspension (over 10 g of mud per liter of water) that it is given a special name, "fluid mud," by scientists. Based on our field work, it seems that the southernmost channel (nearest to Australia and the Great Barrier Reef) is currently the most active. We also concluded that tides and waves working together may allow fluid mud to be leaking out of the Delta and into the marine environment, possibly taking some Ok Tedi mine waste along with it.[23]

Part of our work included taking grab samples of the surface sediments, and to do this, I had borrowed an expensive Smith McIntyre grab from Phil Gibbs in Sydney. We collected over 100 samples in the Delta using the Smith McIntyre grab. At the last station, wouldn't you know it, but the ship ran aground on a sandbank! Never mind, we could still collect a grab sample even though we were aground, so over the side, the grab went, but when it was brought on deck, the stainless steel pan that collected the sample had fallen out. I had lost a vital part of my colleagues sampling machine! And I knew it must be right there, invisible in the murky water, sitting in the mud next to the ship.

Without thinking, I kicked off my deck boots and donned mask, snorkel, and fins to try and locate the missing pan. Into the water and down I went. It was only about 3 m (10 feet) deep, but the water was all inky darkness. There was zero visibility in that muddy water. Totally blind, I groped around on the bottom, feeling for the pan, but in the strong current and black water, I quickly lost my sense of direction. After a few attempts, I had to admit defeat. The pan was lost!

Back on deck I shared my experience with the team, and we shook our heads at the misfortune of losing the valuable sampling pan. Then one of the crew mentioned something I hadn't taken into account. "Good thing there were no crocs in the water when you were in there!" he said.[24] Seeing the stunned look on my face, another crew member said "Don't need to worry about crocs in here mate. The sharks got 'em all!"

Today, the Fly River Delta remains a wilderness area, but its environmental integrity has been compromised by the Ok Tedi mining disaster. The Fly River could take 200 years to recover. The long-term consequences of the disposal of over 2.7 billion tons of mine waste into the Fly River Delta over the last 35 years is unclear – but there can be no doubt that there will be consequences.

---

[23] Harris et al. (2004).

[24] The largest confirmed saltwater crocodile ever recorded was from the Fly River Delta. Its dried skin was 6.3 m (20 feet 8 inches) long.

Chapter 9
The Face of the Deep

"In addition to a gentle, voluminous, and un-relenting snowfall of clay and ooze the ocean floor intermittently receives consignments of a large size: rocks plucked from the land and later dropped from melting ice- bergs; stomach stones disgorged by sea lions; rocks borne to sea by rotting giant algae; garbage, paper, radioactively contaminated tools and waste; junk, wrecks from peace and war; artillery projectiles; clinkers from coal- burning steamships; ballast; telephone and telegraph cables; beer bottles, tin cans and much more, imaginable, virtually unimaginable, and sometimes, unmentionable."
Bruce Heezen and Charles D. Hollister
The Face of the Deep, 1971

Abstract Still on board our hover car, in this chapter we will explore the deep ocean's submarine canyons, seamounts, plateaus, deep ocean trenches, and the mid-ocean ridge. We will meet pioneer marine geologist Francis Shepard and learn how submarine canyons are formed by underwater landslides. We will learn that there is a connection between the testing of nuclear bombs in the Pacific and Charles Darwin's theory for the formation of coral atolls. We will learn about Zealandia, the 8th undiscovered continent. We will ponder what creatures inhabit deep-sea sediments leaving only traces in the mud of their existence and what happens to its body when a whale dies. Will humans mine manganese nodule deposits on the deep ocean floor, knowing that once destroyed this habitat will never recover? To mine or not to mine – it is a moral dilemma.

Keywords Continental slope · Submarine canyon · Zhemchug Canyon · Francis Shepard · Turbidity current · Cold-water corals · Trench · Hadal zone · Red clay · Seamounts · Darwin's atoll theory · Guyots · Atlantis · Submarine plateau · Carbonate compensation depth · Manganese nodules · Hydrothermal vents

© Springer Nature Switzerland AG 2020   143
P. T. Harris, *Mysterious Ocean*, https://doi.org/10.1007/978-3-030-15632-9_9

Our hover car is now parked on the edge of the continental shelf, and we are looking out over the edge where the seabed slopes away from us into the distance. Behind us is the "algae line" below which no benthic plants or photosynthesizing organisms live; all life on the seabed before us is animal life that depends on the overlying water for food, mainly in the form of organic detritus (dead plankton and fish poo) sinking slowly through the water column. The ground behind us is continental crust, and before us are the great ocean basins underlain by oceanic crust composed of basalt. Somewhere down the slope before us is where ocean crust abuts the continental crust. This "continental slope" marks the transition from continents to oceans. And it is here that we encounter one of the most dramatic seafloor features occurring in the oceans: submarine canyons (Fig. 8.1).

∗∗∗

The Grand Canyon in the United States is 446 km (277 miles) long, up to 29 km (18 miles) wide, and it has a mean depth of over 1800 m (5900 feet). But submarine canyons incised into the continental slope are much larger. The largest known is the Zhemchug Canyon located in the Bering Sea, off the coast of Alaska (Fig. 9.1); it is over 400 km (250 miles) long and 100 km (63 miles) wide, and it is incised with a vertical relief of around 2600 m (8500 feet). The longest submarine canyon in the world is the Bengal Canyon off the coast of Bangladesh; it is not as deeply incised or as wide as the Zhemchug Canyon, but it extends for almost 2800 km (1740 miles) across the Bengal submarine fan complex into the Bay of Bengal. There are around 10,000 submarine canyons in the ocean, and they have an average depth of incision of about 2300 m (7500 feet); in other words, the *average* submarine canyon is more deeply incised than the Grand Canyon.

The existence of submarine canyons has been known for over 150 years, since the first systematic mapping of the seafloor began to create reliable nautical charts. But the undisputed pioneer in submarine canyon research is Francis Parker Shepard (1897–1985), who published a textbook in 1948 titled *Submarine Geology*[1]. His textbook, more than any other, established the field now called "marine geology." By as early as 1938, Shepard had compiled the first global database on submarine canyons. This was followed by dozens of papers and books published over the next 40 years to reveal facts and figures that have framed discussions about the origins and evolution of submarine canyons, ever since.

Francis Shepard was born in Massachusetts, studied geology at Harvard, and took his PhD at the University of Chicago. Shepard learned how to sail from his father, owner of the Shepard Steamship Company, and became interested in marine geology while exploring submarine canyons off New England from aboard his father's sailing yachts. Shepard recognized the geological significance of information contained on nautical charts. In 1923, he obtained a global set of charts, whose bathymetric contours depicted features such as glacial troughs and submarine canyons and whose notations indicated patterns in shelf sediments that were contrary to the scientific beliefs of the time.

For shelf sediments, the dogma up until the 1950s was the concept of a graded shelf, whereby sediment close to the shore is sandy because of the strong wave

---

[1] Shepard (1948).

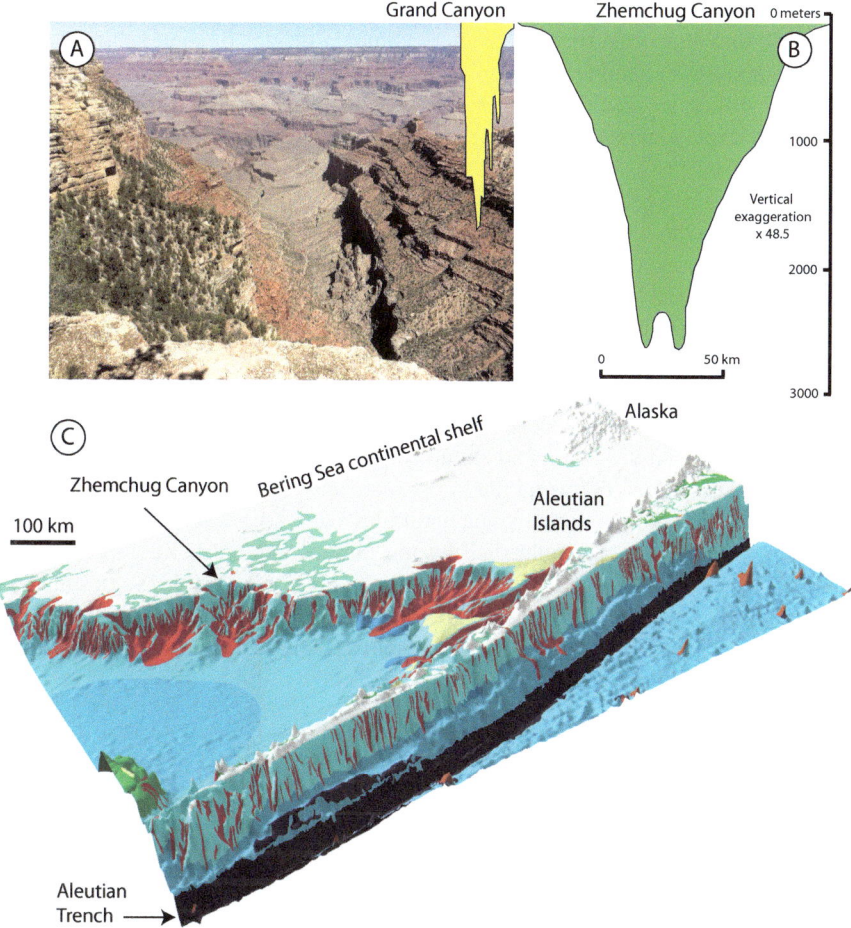

Fig. 9.1 Comparison of the Grand Canyon with the Zhemchug (submarine) Canyon located in the Bering Sea off Alaska. (**a**) Photograph of the Grand Canyon taken by the author from the South Rim, May, 2018; (**b**) cross-sectional profiles (taken from Normark and Carlson 2003) illustrating the gigantic size of submarine canyons compared with their terrestrial cousins; (**c**) three-dimensional map showing part of the Bering Sea with location of Zhemchug Canyon. Submarine canyons are highlighted in red (see Fig. 9.2 for geomorphic feature color codes)

action, but as one moved into deeper water, sediments become muddy and eventually very fine clay, as the effects of wave energy decreased. This is a very logical explanation and is no doubt true in some places. However, Shepard observed that the exact opposite pattern often occurs, with muddy sediments along the coast and sand and gravel found offshore. The explanation is that the distribution of shelf sediments is often times controlled by processes other than wave action. In some areas, the effects of past glaciations control what is found on the seabed (like they do in Puget Sound) and in other places tidal currents play a major role (like they do in the

Bristol Channel). Shepard recognized these factors and collected data from around the world that supported the idea of more complex origins of shelf sediments and the development of continental margins.

The first theories as to how submarine canyons were formed typically involved erosion by rivers making a valley that was later submerged by rising sea level. Early surveys tracked the canyons at their shallow ends (near the coast) which could be explained by a rise of sea level of a few hundred meters, and this seemed feasible and consistent with what was known at the time. But as more and more data became available, it became clear that the largest canyons extended to depths of thousands of meters below sea level. How could sea level have been lowered so far, that rivers could erode valleys thousands of meters deep into the continental slope? Mechanisms other than river erosion must explain canyon formation to account for the global distribution of canyons that currently exist on every continental margin, incised thousands of meters into the slope.

The idea that mass wasting and slumps of rock and sediment on the continental margin could result in dense mixtures of water and sediment (turbidity currents) cascading down the continental slope and miles out to sea was first formally proposed in a scientific paper published by the famous geologist, Reginald Daly in 1936. Daly's idea of the formation of canyons by erosion by turbidity currents was at first rejected by Shepard until compelling evidence of the erosional power of turbidity currents came to light from an unexpected source: the failure of transatlantic telecommunications cables.

***

October 29, 1929 (Black Tuesday), was not a good day for the world's stock markets. Another event occurred 3 weeks later, now long forgotten by the general public, but which subsequently proved very useful to marine geologists. On November 18th of that fateful year, an earthquake occurred on the Grand Banks off Newfoundland, Canada, which caused a small tsunami in Canada and the failure of several transatlantic submarine communication cables that lay on the seabed. It is the submarine cables that are important to this story.

There were several cables that survived the event, so not all communication was lost between Europe and North America, but several of the cables laying in deeper water, off the continental shelf, were broken. Bruce Heezen and Maurice Ewing noticed that six cables actually broke in order of increasing water depth, and they published a paper in 1952 that provided a plausible explanation: a massive turbidity current that originated on the upper slope travelled downhill breaking the cables one by one. By noting the exact time that each cable broke, they estimated that the turbidity current had travelled at a speed of about 50 miles/hour (the speed has subsequently been revised down to at least 42 miles/hour or 19 m/s – still pretty fast!).

In the following decades, several research cruises mapped and sampled the site of the 1929 Grand Banks turbidity current, and a number of scientific papers were published. It has become clear that the 1929 turbidity current was a huge event. The flow of mixed water and mud was several hundred meters thick, and it transported over 150 cubic kilometers (36 cubic miles) of sediment. Mud and gravel, sourced from several tributaries of a moderate-sized submarine canyon called the Saint Pierre Valley, formed the bulk of the turbidity flow. Following the earthquake, the turbidity current flowed for a period of around 11 hours down the continental slope.[2]

Gradually, marine geologists have recognized the significance of this event, and it has transformed our understanding of deep-sea processes and the role of erosive turbidity flows in sculpting the seabed down to water depths of thousands of meters. When you consider that major earthquakes greater than magnitude 7 occur more than once per month somewhere on the Earth (a magnitude 8 or greater occurs about once a year), it becomes apparent that turbidity flows like the one that happened on the Grand Banks in 1929 are probably a common occurrence in the global ocean. They go unnoticed by humans since they can't be seen. Failure of submarine cables is one of the only signs we have that they have ever happened. Once believed to be a quiet and tranquil environment, the deep ocean is now known to experience catastrophic landslides and massive influxes of sediment (among other kinds of disturbances like benthic storms and volcanic eruptions). The deep ocean is not so quiet and tranquil after all.

***

The erosive power of turbidity currents explains how submarine canyons are excavated hundreds of meters into the margin at great depths, thousands of meters below sea level. Rivers may have played a part in the creation of the initial shelf valley, carving a shallow channel across the continental shelf during the ice ages when sea level was lower. The sediment load of rivers delivered to the coast provides the raw material for subsequent turbidity flows. But it is the slumping of the margin and associated turbidity currents flowing down canyons and eroding the seabed that are the fundamental processes of submarine canyon evolution.

We now know that there are about 10,000 large submarine canyons in the ocean with 2000 of the largest submarine canyons cutting into the continental shelf (Fig. 9.2). Only around 150 canyons are known to continue across the shelf as sea valleys that join to river systems on land.[3] The largest, shelf-incising canyons are of special ecological significance because they interact with ocean currents forcing the upwelling of cold, nutrient-enriched waters onto the shelf. Upwelling, in turn, creates an area of enhanced primary production which supports an associated ecosystem with an abundance of fish, seabirds, and whales. For this reason, submarine canyons are targeted by the fishing industry as well as by conservation groups determined to preserve these biodiverse environments.

The shallow "heads" of submarine canyons are often marked by steep, vertical to overhanging cliff faces, especially in cases where the canyon is incised into bedrock. These steep rocky canyon walls are host to clusters of deep, cold-water corals growing on the rocky escarpments. From our hover car, we can see that these corals look different from the tropical corals we are used to seeing in the shallow surface waters. But these deep-dwelling, cold-water corals are just as colorful and beautiful

---

[2] Piper et al. (1999).
[3] Harris and Whiteway (2011).

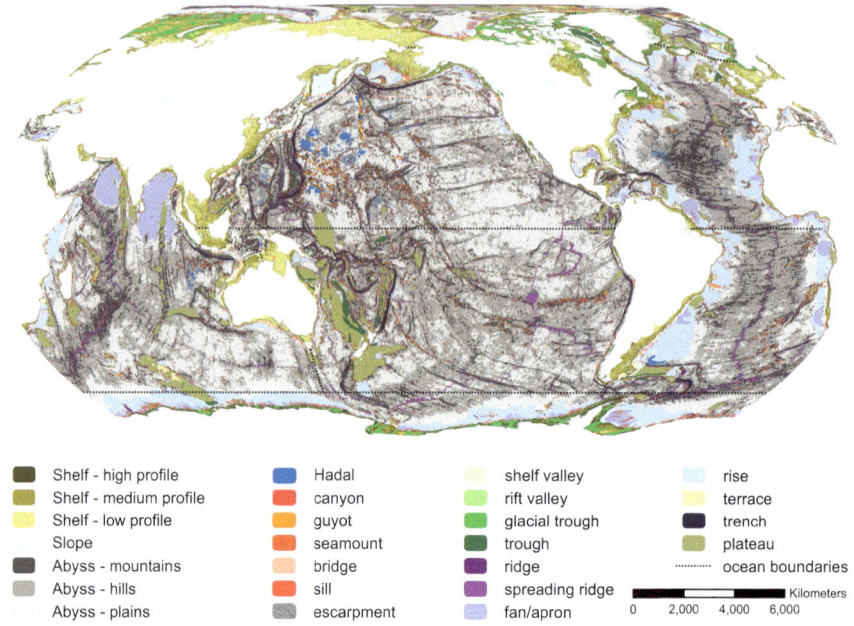

■ Shelf - high profile	■ Hadal	▢ shelf valley	▢ rise
▢ Shelf - medium profile	■ canyon	▢ rift valley	▢ terrace
▢ Shelf - low profile	■ guyot	■ glacial trough	■ trench
Slope	■ seamount	■ trough	▢ plateau
■ Abyss - mountains	▢ bridge	■ ridge	·········· ocean boundaries
▢ Abyss - hills	■ sill	■ spreading ridge	Kilometers
Abyss - plains	■ escarpment	▢ fan/apron	0 2,000 4,000 6,000

Fig. 9.2 Seafloor features of the global ocean. (After Harris et al. 2014)

to look at: red and purple. Cold-water corals also form reefs tens of meters in height and hundreds of meters across. They support abundant fish, octopus, and other animals. Some of the most common cold-water coral species are *Lophelia*, found in the North Atlantic, and *Madrepora oculata* found in all oceans.[4]

Sadly, we can see from our hover car that many of the reefs have been damaged by deepwater fishing trawlers; it may take over 100 years for a trawled cold-water coral reef to recover from the damage caused in a matter of seconds by one pass of a deep trawl.

The heads of shelf-incising submarine canyons are targeted by fishing fleets because of their association with valuable fish stocks. This is the case, for example, off the coast of southern Spain. Intensive fishing here mobilizes bottom sediments at the heads of submarine canyons whenever the fleet is active. A regular turbidity current flows down the La Fonera Canyon from Monday to Friday, created by the bottom trawling activity. The turbidity currents stop every weekend when the fishermen have a day off and leave their boats in the harbor. Trawling not only creates turbidity currents 5 days a week, but the constant plowing of the same area of seafloor, over and over again, has completely reshaped the seabed transforming what was once a complex habitat of ridges and gullies into a dull, smooth, mud-draped wasteland.[5]

---

[4] Hovland (2008).

[5] Puig et al. (2012).

All these bottom trawling and such unnaturally frequent turbidity currents flowing 5 days a week must have an impact on the deep-sea animals and ecosystems. The true impacts may never be fully known.

As we descend down the axis of the deepest part of a canyon, we can see sediment and clumps of seaweed and other debris gently rolling down the slope. These are not turbidity currents but just the normal background activity of the canyon intercepting sediment and organic matter from the shelf and transporting it into the deep ocean. There is two-way traffic here with sediment flowing down the deep canyon axis and water flowing above it, up the canyon causing upwelling at the sea surface.

When turbidity currents do flow, sediment exits the deep ends of canyons, and it spreads out onto the deep abyssal seafloor building a thick, submarine "fan" (named for their fan shape in map view; Fig. 8.1). On passive, Atlantic-type margins, submarine fans extend out across the flat abyssal plain for hundreds of kilometers. Abyssal plains are the flattest areas on earth, extending for thousands of kilometers with virtually no positive topographic relief. The basaltic ocean crust, with its volcanic pinnacles and hills, is buried under more than a thousand meters thickness of mud that has piled up over millions of years. Sinuous channels cut across the fan, like a winding river on land, dispersing sediment far from the canyon mouth.

In some locations, such as around the margin of Antarctica, deepwater currents flowing along the foot of the slope carry sediment away to form giant "drift" deposits. These regularly spaced spurs of mud project outward from the base of the slope, like a series of giant dunes hundreds of meters in height and 10 or more kilometers apart. Their shape is most pronounced adjacent to the continental slope, becoming more rounded and blurred where they merge into the abyssal plain.

Submarine canyons are remarkable places in the oceans, but humanity has not done enough to protect them from the worst impacts of our industries. Of the 15 countries having the greatest numbers of canyons, only 4 have set aside more than 10% of canyon area in marine protected areas (the United States, Australia, New Zealand, and Italy). Several of the biggest canyon-owning countries have not even protected 1% of their canyons, including Canada, Russia, Brazil, and Greece. Protecting canyon habitats from further damage should be a priority for all countries.

***

Twelve men have stepped foot on the moon, and more than 4000 have scaled the summit of Mount Everest, but only 3 people have been to the bottom of the Mariana Trench, the deepest part of the ocean at roughly 11 km (7 miles) deep. On January 23, 1960, the Swiss oceanographer and engineer Jacques Piccard and US Navy Lt. Don Walsh reached the floor of the Challenger Deep, part of the Mariana Trench located in the western North Pacific Ocean, using the bathyscaphe *Trieste*. The depth of the descent was measured at 10,916 m (35,813 feet). At the time, this was believed to be the deepest piece of the ocean floor.

More accurate measurements taken in 2009 from University of Hawaii's RV *Kilo Moana* found the deepest part of the Mariana Trench to be 10,971 m (35,994 feet).[6] On March 26, 2012, James Cameron, in a one-person vertical "torpedo" submarine, the *Deepsea Challenger*, touched the bottom of the Challenger Deep. Cameron's vessel landed close to the same location where Don Walsh and Jacques Piccard had descended in 1960.

The vertical relief of deep ocean trenches exceeds anything comparable found on land. The deepest canyons on land include the Tsangpo Canyon in Tibet which is 5500 m (18,000 feet) deep and the Kali Gandaki Gorge in mid-west Nepal which is 6400 m (21,000 feet) deep. The ocean trenches exceed these depths in many places, and the Mariana Trench is almost twice the vertical relief of these two on-land examples. If Mount Everest were placed at the bottom of the Mariana Trench, there would still be 2 km (1 mile) of water over its summit.

The "hadal" zone in the ocean is defined as seafloor located in water deeper than 6000 m, and globally it covers an area of about 3.4 million square kilometers (1.3 million square miles). The deep ocean (hadal) trenches, which cover an area of about 2 million square kilometers (0.8 million square miles), are one of the least well-studied habitats on Earth. Observations from the *Trieste* and *Deepsea Challenger* expeditions, plus many unmanned camera and dredge surveys, have provided scraps of information on hadal trench biology. Microbial matts, holothurians (sea cucumbers), giant (17 cm long) amphipods, and acorn worms inhabit this dark and forbidding environment.

Nourishment is sourced from whatever organic matter that may fall into the trench from above but also from the Earth's crust. Subduction of the ocean crust releases water that percolates through the rock, releasing sulfur, methane, and hydrogen which are food for bacteria. Trenches and troughs are the recycling centers of ocean basins, where ancient ocean crust is subducted into the mantle. Here it is sequentially melted to distill out granite and continental crust (as discussed in Chap. 2). The remainder is melted and blended in the mantle, only to re-emerge in the rift valley of the mid-ocean ridge after many millions of years. In addition to ocean crust, all the sediment that falls into the trench from land (transported down submarine canyons), together with sediment and fossils riding passively on the ocean plate, is conveyed downward into the mantle for recycling.

Viewed from our hover car, the ocean trench terrain is very rocky and disturbed by regular avalanches of rocks and mud sliding in from the steep side walls. In places, masses of rock and sediment have slumped into the trench, nearly filling it and making a bridge from one side of the trench to the other. Such bridges are common features in ocean trenches, acting as partitions that divide the trench into segments while also providing a handy connection across for any would-be pedestrians.

Deep ocean trenches are the signature geomorphic feature of Pacific-type continental margins and ocean basins. The North and South Pacific together contain 75%

---

[6] https://archive.is/20120524194643/http://www.soest.hawaii.edu/UMC/Reports/Archives/KMreportJuneJuly2009.html

of all the ocean trenches. The other oceans share the remaining 25% of trench area among them; the next largest is the Indian Ocean with 8% (including the Java Trench) followed by the North Atlantic with 6% (including the Puerto Rico Trench). Apart from the Arctic Ocean, there are small areas of trench in every ocean basin, including the Mediterranean Sea.

At the top of the trench, looking seaward with our backs to the continent, we see before us the vast abyssal seafloor. Trenches mark the transition from granitic, continental crust, on the continental slope side, to basaltic ocean crust on the abyssal seafloor side. On the continental slope side, the trench wall is draped with sediment coming from the land mixed with marine sediments (shells from marine plants and animals). On the oceanic side, the trench wall is composed of basaltic rock which is thinly draped by layers of sediment that has fallen through the water column. This sediment, called "red clay," is made of dust blown over the ocean from volcanic eruptions and deserts that have settled to the ocean floor.

The North Pacific has the largest area of hadal ocean floor, that is, the area deeper than 6000 m (20,000 feet). In this region, located toward the end of the great ocean conveyor, respiring animals have used up most of the dissolved oxygen and released carbon dioxide into the water column. The intense pressure of over 600 atmospheres combined with high carbon dioxide levels makes the water corrosive. It dissolves away calcareous shells, and all that remains is "red clay" with some organic plankton (silica) shells and fish's teeth (teeth are made of the mineral apatite which dissolves very slowly).

One of the uses humans have considered for ocean trenches is as a receptacle for toxic waste, particularly radioactive waste from nuclear power plants. The idea is that the waste will be subducted into the mantle and melted, rendering it harmless. The main problem with this idea is that the rate of radioactive decay of nuclear waste (about 10,000 years) is very fast compared with the rate of subduction (around 5–8 cm/year or about 4 inches/year). In 10,000 years the waste will have moved a mere 800 m (2400 feet) at most. Meanwhile, radiation will have been released into the water moving through the trench and been taken up by trench biota, entering the food web. Using ocean trenches for disposal of nuclear waste is one idea that well and truly needs to be buried (preferably deep on tectonically stable land).

***

In Jules Verne's science fiction novel *Mysterious Island*, a group of escaped US Civil War prisoners are marooned on a volcanic island. The novel ends with a volcanic eruption that destroys the island which then sinks into the ocean just as the castaways are rescued. Verne's imaginative description of the islands' cataclysmic ending is actually quite an accurate description of an explosive volcanic eruption, involving seawater rapidly flooding into a magma chamber and being converted into steam. The books' publication in 1864 preceded the eruption of Krakatoa in 1883, which erupted exactly as described in *Mysterious Island*, with uncanny realism that demonstrated Verne's excellent understanding of the geological processes involved.

Volcanic islands are in fact the emergent summits of submarine volcanos. Their submerged cousins are called "seamounts." Both volcanic islands and seamounts are submarine volcanos that have formed over a hot spot in the mantle, along the mid-ocean spreading ridge or at some other location of volcanic activity. Rarely, the seamount may rise thousands of meters in height above the seafloor until it emerges above sea level as a volcanic island. There are not that many active volcanic islands in the ocean, comprising perhaps around 30 or so island groups (like the Hawaiian Islands). It appears that less than half of 1% of seamounts make it to the surface to form an island – in other words, there are 200 seamounts in the ocean for every volcanic island.

The eruptions of lava on the island of Hawaii give us some idea of how seamounts form and grow. Since their eruptions occur mainly underwater, we can deduce that seamounts are made of pillow basalts, like those seen in underwater videos from Hawaii. Pillow basalts are essentially large blobs of lava that have been squirted like toothpaste from a lava tube. The red-hot lava cools rapidly in contact with seawater, and the pillow-shaped blob of lava breaks off from the tube and tumbles down the flank of the volcano. A seamount may be simply a cone-shaped pile of pillow basalts on the seabed with an internal plumbing system of lava tubes and dykes.

Seamounts are truly massive piles of pillow basalt – to qualify for the label of "seamount" according to the International Hydrographic Organization, the volcano must reach an elevation of at least 1000 m (3280 feet) from base to summit. The average seamount has a life that begins with a hot spot in the mantle piercing through a thin spot in the ocean crust to cause a volcanic eruption on the seabed. The pillow lavas pile up making a cone-shaped volcano. Lava makes its way upward, and eruptions continue at the summit making pillow lavas that tumble down the slope piling up one by one. If the volcano remains active for long enough, it can build up a pile that is over 1000 m (3280 feet) tall and typically around 30 km wide (the average seamount is around 800 square kilometers or 312 square miles, in area). The life span of an active seamount is typically between 1.5 and 10 million years before all volcanic activity ceases. The seamount then slowly subsides as the crust cools around it, and its shape is masked by a thick draping of sediments.

The quiet, slowly flowing lava of Hawaiian-type volcanos is quite different from the explosive eruption of Krakatoa in 1883. The reason is because of the composition of the lava; basaltic lava sourced from ocean crust results in sluggish, Hawaiian-type volcanos, whereas a mixture of ocean crust with continental crust containing silicate minerals gives the more explosive, Krakatoa-type. Since most seamounts erupt in water that is thousands of meters deep, the extreme pressure at abyssal depths muffles the eruption so that there is hardly any sign of it at the sea surface. The submarine eruption of one volcano located north of New Zealand was only noticed because of large rafts of floating pumice seen by a curious geologist who happened by on a passenger plane.[7] There are thousands of volcanos in the ocean, and we can only wonder at how many underwater eruptions have occurred that we are unaware of? It is a mystery!

[7]Carey et al. (2014).

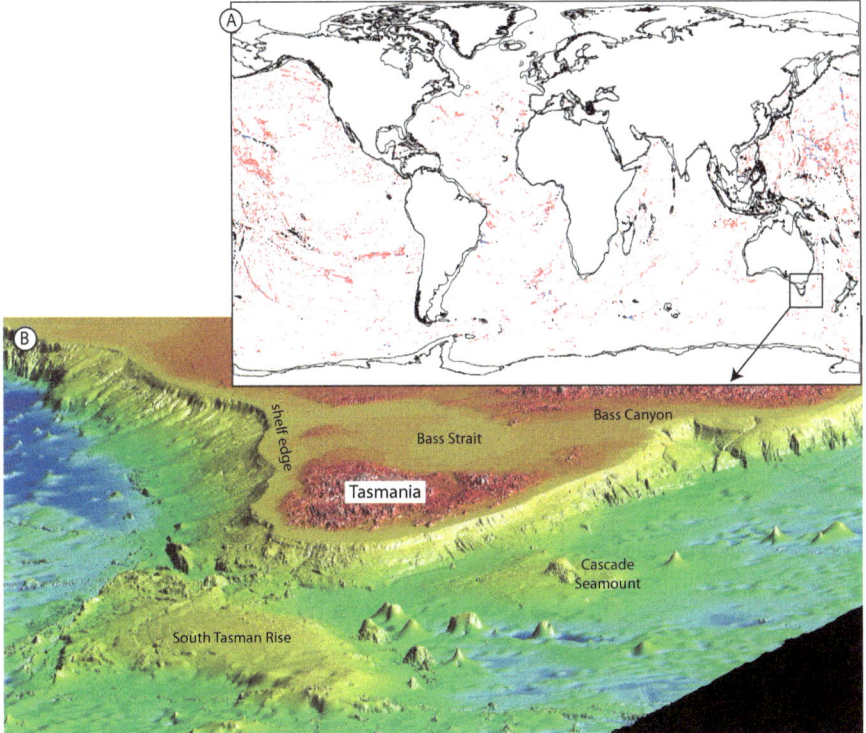

Fig. 9.3 (**a**) global distribution of seamounts, extracted from Harris et al. (2014), showing seamounts (in red) and guyots (in blue); (**b**) three-dimensional bathymetric model of the seafloor off southeastern Australia, showing several seamounts including the Cascade Seamount, the South Tasman Rise submarine plateau, and the Bass Canyon

***

What is the tallest mountain on earth? The answer is that it depends on what you choose as your reference level. Mount Everest rises 8848 m (29,028 feet) above sea level, but the surrounding terrain of the Himalayas is already at a high altitude. In comparison, the summit of Mauna Kea on the island of Hawaii is 4207 m (13,802 feet) above sea level, but the summit is actually 10,203 m (33,474 feet) above the level of the surrounding ocean floor. The island of Hawaii has a basal area (seafloor footprint) of around 11,000 square kilometers (4290 square miles). It can be argued that, depending on your point of reference, Hawaii is taller than Mount Everest. From any perspective, seamounts and oceanic volcanos are gigantic features of the Earth's surface.

There are about 10,000 true seamounts in the global ocean (Fig. 9.3) with hundreds of thousands of smaller volcanic peaks, knolls, and hills that are less than 1000 m in elevation (*wanna-be* seamounts). There are also around 3800 ridges on the seafloor that are over 1000 m in elevation (Fig. 9.2). Ridges can be volcanic in origin but are most often produced by fracturing and buckling of ocean crust associated with seafloor spreading.

Upwelling of lava from the mantle is common along the mid-ocean spreading ridges, but there are other mantle hot spots distributed randomly across the ocean floor. Some hot spots remain active for periods of time that are long compared with the life span of a seamount (multiples of 10 million years). While the hot spot remains at a fixed location, ocean crust is moving past driven by plate tectonics. As the ocean crust passes by, one volcano is formed after another to make a chain of seamounts – for example, Hawaii is the above-surface part of the Emperor seamount chain. There are many seamount chains in the ocean, but the majority of seamounts appear to be more or less randomly distributed. Even so, there are important differences between the oceans in terms of the numbers of seamounts they contain.

Of the world's 10,000 seamounts, 7000 are in the Pacific Ocean, and only 1700 are in the Atlantic, while the Arctic Ocean has less than 30 seamounts.

Why is this so?

The reason is because sedimentation in the Atlantic, Arctic, and Southern Oceans has buried many seamounts so that only their summits are visible (if they can be seen at all). Remember that sediments entering the Pacific are captured by trenches and troughs, whereas sediments pile up in thick layers next to continents in the Atlantic, Arctic, and Southern Oceans. The Atlantic *appears* to have fewer seamounts than the Pacific simply because many Atlantic seamounts have been smothered and buried under thick layers of sediment.

***

Our knowledge of seamounts started with studies of volcanic islands that were visited by explorers and early scientists. Charles Darwin noticed that there were different patterns of coral reef growth around volcanic islands that he visited during his voyage aboard HMS *Beagle* during 1831–1836. Reefs appeared to grow close to shore around the tallest (sometimes active) volcanic island peaks. On other islands, the volcanic peak seemed eroded and volcanically inactive, and around these islands, coral reefs were seen to be growing further offshore, with a proportionately larger reef lagoon behind. Then there are oval-shaped coral atolls comprised of coral reefs growing around a central shallow lagoon without any sign of a volcanic peak.

Darwin wondered if there was somehow a connection between volcanic islands and coral atolls he had seen? He deduced that he was seeing volcanic islands and their surrounding reefs at different stages of growth and development. Reefs first colonized the shore around young volcanic islands. Once active volcanism ceased, the volcano subsided, and the peak was eroded by rain and ocean waves. The reefs, meanwhile, maintained their position close to sea level around the island's original perimeter so that, as the volcano sank, the reefs appeared to be growing further offshore. Sediment derived from the reefs and erosion of the volcano washed into the lagoon to maintain its shallow depth. Eventually, the central volcanic peak eroded away by rain and wave action until all that remained is an oval-shaped coral reef. And the end result? An atoll! This idea is now known as Darwin's atoll theory, published in his book *The Structure and Distribution of Coral Reefs* in May 1842.

Darwin's atoll theory triggered a scientific debate over the origin of atolls that would take a century to finally resolve. Were atolls really extinct volcanos with corals on top? Or are they actually coral reefs through and through, growing upward from abyssal depths? Darwin suggested that drilling into coral islands would prove the point; he predicted that drilling down a few hundred meters below the sea surface, one would penetrate through reef limestone and then encounter the volcanic basaltic rock basement upon which the original reef grew. Charles Darwin died in 1882, and he did not live to see his atoll theory put to the test.

Several reef drilling expeditions were mounted to test Darwin's theory. In 1896 the Royal Society awarded the Australian geologist from the University of Sydney, T. W. Edgeworth David, a grant to carry out a drilling expedition to Funafuti Atoll in Tuvalu. The first drill hole he attempted in 1896 only reached 30 m (98 feet) depth. The next year Edgeworth David drilled to 210 m (690 feet), finding only coral limestone, and in 1898 a depth of 340 m (1115 feet) was reached, the limit of drilling technology at that time, but still in coral limestone. It seemed that Darwin had either underestimated the thickness of atoll limestone or that his theory was incorrect.

Fifty years went by without any further evidence coming to light. Then the US Geological Survey, in support of nuclear bomb testing in the Pacific, was given the task of drilling into atolls of the Marshall Islands prior to the detonation of atomic bombs. Testing of nuclear bombs on coral atolls took place in several locations. The United States detonated 105 atmospheric and underwater nuclear bombs between 1947 and 1963 at Bikini Atoll, Enewetak Atoll, and Johnston Atoll. The French carried out around 180 nuclear tests at Mururoa Atoll between 1966 and 1996, 41 of them were atmospheric tests above the atoll and the rest below ground in holes drilled into the limestone. The explosions caused the Atoll to crack apart, and large blocks of rock slumped off the Atoll's flanks. The British detonated eight nuclear bombs in the atmosphere above Malden Island and Christmas Island in Kiribati between 1957 and 1958.

One positive result of all this nuclear testing was the geologic drilling carried out to better understand the composition and structure of atolls. Drilling by the USGS on Enewetak Atoll in 1952 reached a depth of 1411 m (4629 feet). This was the first atoll drill site to reach basement rock beneath limestone, proving that the foundation of Enewetak Atoll is a basaltic volcano, just as Darwin had predicted!

A second line of evidence in support of Darwin's atoll theory came from seabed mapping. Bathymetric maps in the 1800s were crude representations at best of the true complexity of the seafloor. The advent of seafloor mapping using sonar in the early twentieth century provided the marine geologist Harry Hess (of seafloor spreading fame) with the perfect distraction for his naval career during World War 2. A peacetime Professor of Geology at Princeton, Hess joined the US Navy during the war in command of an attack transport ship equipped with sonar. Hess kept the sonar running during operations in the Marianas, Philippines, and Iwo Jima. During the long transit voyages, he noted strange, flat-topped seamounts (table-mounts), which he called "guyots," named for the flat-roof shape of the geology building at Princeton, Guyot Hall (which itself is named for the Swiss-American geologist, Arnold Henry Guyot).

Hess published his paper "Drowned ancient islands of the Pacific Basin" in 1946 in which he described guyots. The flat top of guyots was attributed by Hess to wave erosion of a seamount, while its summit was at sea level, which is correct in some cases. But many guyots, possibly the majority of the largest ones, are drowned coral atolls as shown from samples of coral dredged from their summits. Guyots are the evolutionary end products of Darwin's atoll theory, although their existence was unknown to him at the time.

***

Seamounts are biological oases in an oceanic desert. Due to clear water conditions that often occur in the open ocean over seamounts, photosynthesis is possible at great depths. For example, live coralline algae were recovered at a depth of 268 m (879 feet) from the San Salvador Seamount in the Caribbean Sea. Seamounts may generate coastally trapped waves and internal waves and may amplify the ocean tide. Mixing in the waters over a seamount can therefore be 100–10,000 times greater than mixing rates in the surrounding ocean. The resulting strong currents carry food to filter feeders, remove waste, and inhibit sedimentation, perfect habitat for filter feeders like cold-water corals, gorgonians, actinarians, pennatulids, and hydroids. The collective biodiversity of seamounts is immense – it is an amazing fact that more species of corals, for example, have been described by scientists from the deep ocean than from shallow-water, tropical coral reefs.[8]

Seamounts rising to within a few hundred meters of the sea surface interact with waves and surface currents, deflecting flow and forming local eddies. There are around 1500 seamounts with summits at a depth of less than 1000 m. In some cases, an eddy may become trapped over the seamount to form a closed circulation cell known as a "Taylor column." These oceanographic features have been observed to persist over an individual seamount for over 6 weeks, and the associated turbulence may induce upwelling and locally enhanced primary productivity in an otherwise oligotrophic (biological desert) oceanic regime.

Enhanced primary productivity over seamounts attracts migratory birds and pelagic species, such as sharks, tuna, billfishes, and cetaceans. The relatively abundant benthos attracts and supports a range of demersal and epipelagic fish species and makes seamounts the preferred aggregation and spawning grounds for deep-sea species such as orange roughy, black cardinalfish, pelagic armorhead, and alfonsino. The abundance of life is as much as 100 times greater on seamounts than occurs on the adjacent sediment-covered abyssal plain. Amidst the vastness of the oceans, seamounts are indeed oases for a diversity and abundance of marine life.

The relative abundance of life on seamounts attracts the attention of fishermen interested in harvesting their fish and precious corals (where any resources is exploited in an unsustainable manner, "harvesting" is a euphemism more aptly described as "mining"). Seamount fisheries are inherently unsustainable for many reasons. The longevity of deep-sea fish (e.g., orange roughy are estimated to live between 80 and 100 years, and they do not reach reproductive maturity until they

---

[8]Cairns (2007).

are 30) and the relatively small size of their isolated populations make them very vulnerable to overexploitation. As the seamounts have been located and fished, one by one, their populations are rapidly decimated; the fishery booms and busts in around 10 years. From 1990 to 2005, seamounts yielded about 100,000 tons of fish per year. Most seamount fisheries today are either in decline or have been abandoned because they are no longer economic.[9]

Precious corals are even more vulnerable than seamount fisheries. It could take centuries for corals dredged from a seamount to recover. The damage caused by seamount fisheries is unknown (scientists have no "before" and "after" surveys to go on), but comparisons of trawled and un-trawled seamounts indicate that the effects are devastating. The biology of seamounts is very poorly studied, and it is a tragedy to think that species may already be extinct before we even knew they existed. This is not just a moral dilemma surrounding the loss of species. Consider, for example, their potential value to medicine as new sources of pharmaceuticals.

Around 60% of seamounts occur on the high seas and thus fall outside the jurisdiction of any country. Fishing here is unregulated. Only large, deep-sea trawlers can access these remote locations, and it is ironic that only government fuel and other subsidies have made it economically feasible to fish them; government subsidies are estimated to equal 25% of the value of deep-sea trawl fisheries and substantially more than their net profit. In effect, governments are underwriting the destruction of seamount fish stocks and deepwater coral ecosystems. Even though the high seas are the common heritage of people living in all countries, ten nations capture 71% of the landed value of catches in international waters – put simply, the benefits flowing from "mining" the high seas fish stocks are not being shared equally among all nations.

A possible solution, supported by many scientists and conservation groups, is to simply ban all fishing on the high seas. The high seas provide less than 5% of the global, wild fish captured each year – it would not be a significant loss to world food supplies – but it is a very destructive fishery. Modelling suggests a ban on high seas fishing could actually increase the global wild fish catch by the spillover effect from what would essentially comprise a global marine protected area.[10]

The good news is that raised awareness is increasingly making seamount fishing politically unacceptable. In 2009 the United Nations General Assembly passed a resolution recommending that environmental impact assessments be carried out prior to the development of any new deep-sea fisheries and that steps should be taken, such as creating marine reserves, to ensure the conservation of vulnerable habitats. These are positive developments, but there is still a long way to go to fully protect seamounts and before we will begin to truly understand their unique ecosystems.

***

Plato created the myth of a lost continent, Atlantis, in his story of how ancient Athens defeated the Atlanteans, thereby demonstrating the superiority of his

---

[9] Koslow et al. (2015).
[10] Sumaila et al. (2015).

concept of the "ideal state." Atlantis was of minor importance to Plato, but it has spawned scores of fictional books, music, films, and poems that have embellished the myth into a metaphor for something that can no longer be obtained. The idea of a lost continent (or lost continents) has appeal on a philosophical level, but scientifically speaking there is an element of truth in the myth that Plato could not have expected. There are, in fact, many micro-continents submerged ("lost") beneath the ocean.

According to the theory of plate tectonics, continents spilt apart and fuse together again as they float passively on the Earth's mantle, transported by mantle convection cells. Continental fission and fusion does not always occur neatly, and there are stray, orphaned bits of continents that break off and become separated from their larger parents during the rifting process. The disintegration of Pangea 200 million years ago created not only the six large continents but also a number of smaller continental fragments that are scattered around the globe. Some fragments have remained above sea level, like the islands of Madagascar, Timor, and New Zealand. Others sank beneath the ocean surface. From a seabed geomorphology perspective, these sunken continental fragments are known as submarine plateaus.

It is estimated that there are over 180 plateaus in the modern ocean (180 Atlantises!). They cover about 5% of the ocean floor and have an average size of around 100,000 square kilometers.

Submarine plateaus are typically submerged 1000–2000 m below sea level, rising above the 4 km deep (abyssal) oceanic crust that surrounds them. Their origins are sometimes as continental fragments, sometimes as large volcanic features, and sometimes a combination of both. The Kerguelen Plateau in the southern Indian Ocean and the Ontong Java Plateau in the western equatorial Pacific are examples of "large igneous provinces," whose origin is primarily volcanic. These features formed over millions of years, built up layer upon layer, from repeated volcanic eruptions. The Seychelles-Mauritius Plateau (also called the Mascarene Plateau) in the Indian Ocean is an example of a plateau having a composite origin (continental fragments overprinted with volcanic deposits).

Each micro-continent (plateau) has its own story and different history, depending on when it broke away from Pangea or Gondwanaland (if it is a continental fragment) or when its formative volcanic eruptions occurred (if it is a large igneous province). The history of the Kerguelen Plateau large igneous province, for example, has been revealed by drilling into its thick sediment layers.[11] This plateau is 1.3 million square kilometers (500,000 square miles) in area and is located in the southern Indian Ocean. Fossils and pollen found in sediments indicate that the Kerguelen Plateau was formed around 100 million years ago by massive volcanic eruptions that created a large island. By 50 million years ago, it was covered in lush ferns with a tropical climate. As the crust cooled, the Plateau slowly subsided until it sank below the ocean surface around 20 million years ago, leaving just two islands (Kerguelen Island and Heard Island) perched above sea level on top of the Plateau.

---

[11] Frey et al. (2000).

The world's largest plateau complex is located in the South Pacific Ocean, extending from south of New Zealand to northeast Australia, including Campbell Plateau and Lord Howe Rise. Together with New Zealand, these features comprise the *Zealandia* micro-continent (the 8th continent?), covering an area of nearly 5 million square kilometers (2 million square miles). For comparison, the area of Greenland, the world's largest island, has an area of about 2.2 million square kilometers (860,000 square miles). Australia, the world's smallest continent, has a land area of around 7.7 million square kilometers (3 million square miles), which increases to around 12 million square kilometers once you add in the continental shelves and include Papua New Guinea (which is actually part of the Australian continent).

You may recall from Chap. 3 that the freeboard (mean elevation) of a continent is directly proportional to its size, due to the amount of heat from the Earth's mantle that they trap beneath them. Some continental fragments (plateaus) are too small to trap much mantle heat, and therefore they do not rise above sea level. The rule of thumb is that the smaller the plateau, the deeper it will rest beneath the ocean surface. If Zealandia is so large, why is it submerged below sea level?

The answer is because Zealandia is comprised of several separate pieces of crust that each split off from larger continents. The pieces are adjacent to each other, but they are not fused into a single continental block. Therefore, they do not behave as a single entity in trapping mantle heat. The history of one of the largest pieces, Lord Howe Rise, serves as an example.

In October, 2007, I was the survey leader on an expedition to map and collect seabed samples from the Lord Howe Rise in the Tasman Sea (Fig. 9.3). Our aim was to map the northeastern part of the Rise as part of Australia's frontier petroleum exploration program. We were tasked with providing an assessment of the seabed and its environmental assets, such as notable geological features and biodiversity values. Geoscience Australia had leased the New Zealand vessel RV *Tangaroa* for this survey.

Lord Howe Rise is a fragment of continental crust that separated from Australia between 50 and 80 million years ago. The eastern margin of Australia was joined to the western edge of Lord Howe Rise at that time. New Zealand split away from Gondwana about 85 million years ago, from what is now Tasmania and Antarctica. A rift valley formed and began the separation of Lord Howe Rise from Australia during the Cretaceous; 100-million-year-old volcanic rocks dredged from the Rise are from this early phase. Gradually, a seaway was formed between it and Australia, and by around 50 million years ago, Lord Howe Rise was a separate, long, skinny continent. It gradually subsided to become a shallow shelf sea studded with volcanic islands that slowly sank below the ocean surface. By 20 million years ago, it had become submerged to oceanic depths. There is a spreading ridge system running through the Tasman Sea that was formed at the time of the rifting of Lord Howe Rise from Australia, and this became dormant, also at around 20 million years ago.

Today, Lord Howe Rise has an average depth of around 1500 m (4900 feet), with two small volcanic islands, Lord Howe and Ball's Pyramid, rising above sea level.[12]

---

[12] Harris (2011).

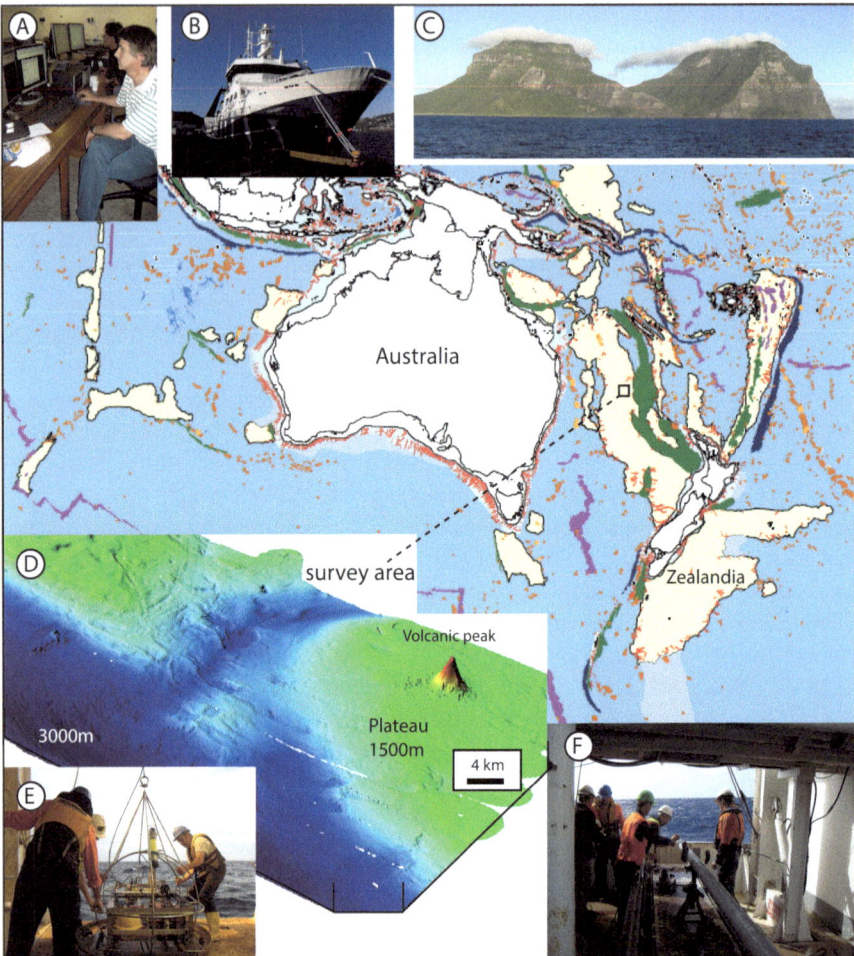

Fig. 9.4 Images from Lord Howe Rise survey 2007: (**a**) Cameron Buchanan monitoring multibeam sonar data acquisition and onboard data processing; (**b**) RV Tangaroa in Wellington Harbour, New Zealand; (**c**) Lord Howe Island; (**d**) multibeam sonar data shown as a three-dimensional view of part of the Lord Howe Rise submarine plateau with a small (about 300 m tall) volcanic peak; (**e**) launching an underwater camera; (**f**) piston core recovered on board. The map in the background highlights the submerged continent "Zealandia," manifest as plateaus (yellow shading) extending north and south from New Zealand. Note the many plateaus around Australia. The small area in the black square was mapped using multibeam sonar from RV *Tangaroa* over a 4-week period in 2007

Our survey aboard *Tangaroa* mapped an area of around 25,000 square kilometers (9750 square miles) using multibeam sonar providing a high-resolution (25 m pixel size) image (Fig. 9.4). We discovered that the Rise is almost everywhere draped with thick deposits of calcareous (made of dead plankton) sediment. A few isolated volcanic peaks provide the only rocky substrate, habitat for sparse bamboo corals, found only on one side of the volcano facing into the prevailing current. They live where currents sweep tiny food particles into their waiting tentacles, making growth rings (like tree rings) that testify to their age of many thousands of years. We found bamboo corals were confined to around 31 square kilometers (12 square miles), equal to about 0.12% of the mapped surface area – hard rocky substrate is very rare on Lord Howe Rise!

Plato's myth of Atlantis, the lost continent, is indeed a geologic possibility. The only trouble is with the timing because it takes millions of years of gradual subsidence to sink even a micro-sized continent. If there ever was a lost continent of Atlantis, it subsided deep below the ocean surface and disappeared long before human beings evolved.

***

Back in our hover car, we leave the submerged plateaus behind us. The view looks similar to a snow-covered terrestrial landscape, with everything covered in a thick blanket of sediment. Sediment is sourced from the surface ocean, above, where planktonic plants and animals live in the photic zone (water depths where sunlight penetrates). Grazing zooplankton (planktonic animals) and fish excrete fecal pellets that transport the sediment into the abyss. The constant "rain" of sediment from above drapes the seafloor like a blanket of snow.

This sediment-draped landscape seems, from a distance, to be flat and featureless. Our view of the seafloor as a monotonous expanse of mud is a matter of perspective. If we descend in our hover car to a few meters height above the seabed, a different picture emerges. Now we can see that the muddy surface is covered with strange marks, tiny footprints, pits, and holes. Some are shaped like stars, and others conjure the spokes of a wagon wheel, while others are elongate grooves leading aimlessly across the mud-scape (Fig. 9.5). What is immediately apparent is the absence of the creatures who must have created these features. What animals made these markings? And where have they gone?

The deep ocean floor is alive with creatures making a living from what has fallen from above. All forms of organic matter, from dead fish to the fecal pellets of plankton, fall from the ocean surface to become food for the deep ocean benthos. This ecosystem is wholly dependent upon the surface ocean for its food supply. The residents here feed in many ways. Some crawl over the muddy surface, munching as the go. Others live within the sediment (creatures known as "infauna" since they live within the sediment), burrowing through the detritus in search of a tasty morsel. Still others make a burrow and only rarely venture out to feed, reaching out from their home with an arm or tentacle.

Now the markings in the mud can be understood. They are tracks and trails made by the different feeding habits of the animals hiding in the mud. The animals themselves are seldom seen. The need for a hiding place raises its own question; what are these animals hiding from? Clearly there must be fish, squid, or other predators that hunt in the mud for prey. This deep ocean ecosystem is a poorly known part in the ocean, where what is unknown vastly exceeds what is understood by scientists.

Consider, for example, a dead whale.

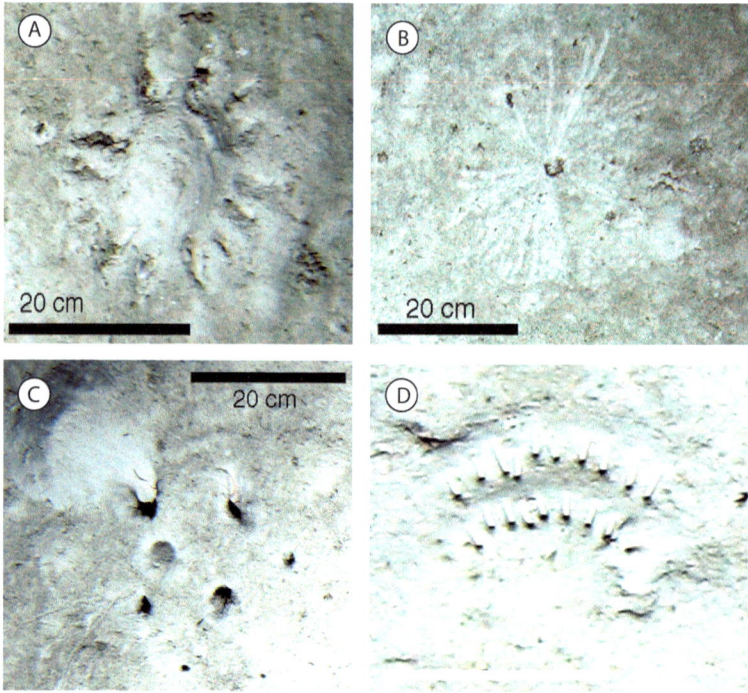

Fig. 9.5 Seafloor photographs from Lord Howe Rise survey 2007: (**a**) radial pattern of depressions with a central sediment mound; (**b**) central burrow with radiating feeding traces; (**c**) burrows with sediment ejected forming a small mound; (**d**) "caterpillar" tracks. The animals that made these marks on the seabed are unknown

When the carcass of a whale sinks to the abyssal seabed, it brings a pulse of organic matter equal to 2000 years' worth of background sedimentation. The sudden delivery of food triggers an amazing reaction. Within hours of its arrival, crabs, fish, and sharks find the carcass and begin to feast. Within about 4 months, the soft tissue is gone and only the bones remain. But these, too, provide habitat and a food source for worms and mollusks. Finally, over a period of years to decades, the bones decay via anaerobic microbial decomposition supporting sulfide-tolerant species.

If dead whales are an important food component for the abyssal ecosystem, what impact has commercial whaling had on the benthos? The answer is, as you might expect, we don't know. It is estimated that the number of whales falling to the seafloor today is approximately one-sixth of the number prior to the year 1800, when industrial whaling commenced. How could an ecosystem adapt to an 83% reduction in food supply? With great difficulty, no doubt! It is highly likely that some whale scavenger specialist species have gone extinct over the last 200 years, but the truth is that we may never know even approximate numbers.[13]

---

[13] Smith and Baco (2003).

As the seafloor drops away, getting deeper and deeper, we notice something happens to the sediment drape. It appears to be getting thinner! We see more and more pinnacles, volcanic cones, and other rocky features poking through the mud. It looks like there is a "snow line" of sediment, thickly draping the shallower features but getting thinner and disappearing altogether as it moves into water deeper down. What could be causing this?

The answer is that, at depths below around 4000 m, sediments dissolve away like melting snow. Most of the planktonic sediment raining down from above, Rachel Carson's "eternal snowfall," is made of calcium carbonate ($CaCO_3$). With increasing depth, seawater becomes gradually more corrosive, a function of pressure and water acidity (pH). Eventually a depth is reached known as the calcium carbonate compensation depth (CCD), and below this depth calcium carbonate dissolves. Where the CCD touches the seafloor, there is a sort of "snow line" for carbonate sediment. Thick calcareous sediment deposits accumulate above the "snow line," but very little sediment that falls below the CCD depth is left behind. Only some fish teeth, diatom shells made of silica, and "red clay" dust, blown over the ocean from distant deserts, remain on the seabed. Like the snow-level on the side of a mountain, the CCD marks the depth below which very little (snow) sediment accumulates.

As we discussed in Chap. 5, humans have increased the concentration of carbon dioxide in the atmosphere, and this is lowering the pH of the oceans, making them more acidic. In the deep ocean, this is causing the CCD to rise by around 1–2 m/year, which means that the "snow line" must also rise, leaving a zone which was previously above the CCD "snow line" but which is now left stranded below it. Carbonate sediments and shells deposited in this zone will begin to slowly dissolve over the next hundreds to thousands of years, acting like an antiacid pill to relieve the oceans' "acid stomach."[14]

At this rate, within 50–100 years, the CCD will have risen by up to 200 m. This zone will be the "front line" for seabed benthic species that will need to adapt to the changing conditions. Animals living here will need to use more energy to build and maintain their carbonate shells and skeletons because the seawater will have become corrosive. The need for more energy means these animals will presumably need more food.

Where will this "front line" occur in the ocean? How large an area previously located below the "snow line" will be located above it by the year 2100? What seafloor habitats and species will be impacted by this change? These are questions that science has yet to answer.

At a depth of over 4000 m (12,000 feet), the view from our hover car changes dramatically. Here, the seafloor appears like a vast plain of small, rounded cobbles.

Welcome to the world of manganese nodules!

These strange, rounded rocks are formed on the seabed by the direct precipitation of metal hydroxides from seawater. They grow incredibly slowly, at rates of millimeters per million years, and they remain unburied because they occur only at

---

[14] Broecker (2003).

great depth, below the CCD. My only firsthand experience with manganese nodules came in 1986 when I was a postdoc aboard the RV *Rig Seismic* collecting seabed samples from a depth of 5000 m on the Perth Abyssal Plain, off the coast of Western Australia. The nodules we dredged were coated in a friable, black manganese crust that came away in your hand. A few tiny corals were attached to the nodules' upper surface.

Manganese nodules generally contain manganese, nickel, copper, and cobalt, although the relative concentrations of the different elements vary from place to place in the ocean. And because of the valuable metals they contain, many millions of dollars have been invested in exploring ways to mine them. Deep ocean mining has not happened so far, mainly because there are other sources of these metals on land where production costs are much cheaper. But there are also good environmental reasons not to mine them.

How would mining manganese nodules impact on the environment? To find the answer to this question, in 1989 a group of German marine scientists started the DISCOL (DISturbance and reCOLonisation) experiment in the Peru Basin of the eastern South Pacific Ocean. An area of seabed was plowed, and manganese nodules were removed over a control area to mimic the impact of mining operations. The site has been revisited, most recently in 2015, to assess the amount of recolonization and recovery of the disturbed area of seabed.

Results show that there hasn't been any recovery.

The disturbed DISCOL area looked exactly the same in 2015 as it had after the area was initially disturbed 26 years earlier in 1989. From what we know about this remote environment, it will never "recover" from the mining of manganese nodules, in the sense that it can return to its natural, pre-mining state. The cost of restoration of deep-sea habitats will greatly exceed that of restoration on land or in coastal settings,[15] making the extraction of minerals there prohibitively expensive, unless, of course, we agree not to restore or rehabilitate the mined areas. Then the environment will be destroyed permanently. Recovery rates are so slow that, on timescales of human life spans, they are essentially nil.

What will be lost from the Earth's biodiversity if mining manganese nodules in the deep ocean proceeds? The answer is we don't know. It is a mystery. But small hints at how little we understand manganese nodule habitats have recently been revealed. A study published in 2016 reported a previously unknown species of octopus that deposited its eggs onto the stems of dead sponges that had grown on manganese nodules in the Peru Basin.[16] Before this study, the species associated with manganese nodules were limited to a few fish species and filter-feeding animals that attached themselves to the rocks. Another intriguing study published in 2018 from the Clarion-Clipperton Zone, located south of Hawaii, reported pits and disturbed seafloor among a manganese nodule area at a depth of over 4200 m. The scientists believe they are possibly the feeding pits created by beaked whales (possibly hunting that newly discovered species of octopus?).

---

[15] Van Dover et al. (2014); see also Van Dover (2011).
[16] Purser et al. (2016).

An intriguing question about manganese nodules is why they exist at all. Recall that nodules accrete very slowly, at rates of millimeters per million years. In comparison, sedimentation rates in nodule areas are about 1 m per million years. These figures suggest that sediments are piling up around the nodules at a rate that is about 1000 times faster than the nodules grow. So how do nodules stay on top of the sediment? Why aren't they simply buried under the sediment? The answer is – we don't know. It is a mystery.

Two ideas may explain how nodules stay on top of the mud and remain on the surface. One is the "Brazil nut" mechanism. If you take a bowl of mixed nuts and give it a shake, the larger, Brazil nuts rise to the top. Given that there are many earthquakes per million years shaking the nodule/sediment "nut bowl," this could explain how nodules float to the top.

Another possible mechanism is that the nodules are disturbed by marine animals, like beaked whales, when they feed.[17] Every time the nodules are nudged aside by a browsing animal, they are lifted back on top of the sediment pile. If this is the case, manganese nodule habitats are the product of a biologically mediated process. Perhaps both processes, earthquake shaking and animal nudging, are working together?

No matter how they form, it is clear that manganese nodules provide a habitat for species and ecosystems that are almost completely unknown to us. It must be questioned if it is wise to begin mining a resource when we don't really understand how it has formed and we don't know how it relates to dependent ecosystems.

***

Leaving the manganese nodules behind us, we speed away in our hover car. Up ahead are the first peaks of the greatest mountain range on Earth, the mid-ocean ridge.

The mid-ocean spreading ridge encircles the Earth along a total length of over 75,000 km (46,000 miles). They are the Earth's largest volcanic system, accounting for 75% of all volcanic activity on the planet. This volcanism is expressed by the central rift valley that occurs along the entire length of the ridge system (Fig. 9.2).

The mid-ocean ridge rift valleys, as first mapped by Marie Tharp whom we met back in Chap. 2, contain erupting lava that forms new ocean crust. The upwelling lava provides a driving force for the plate-tectonic system. It also heats seawater that seeps into the hot rocks where it is chemically transformed before it is ejected via hydrothermal vents. The hot hydrothermal fluid vented into the ocean is rich in iron and other metals.

The existence of such a hot-seawater circulation system was foreseen by geophysicists. But what was not foreseen was that hydrothermal vents found within the rift valleys support life forms totally alien to what we had known before their discovery. Hydrothermal vent ecosystems constitute one of the most amazing biological discoveries of the twentieth century.

---

[17] Marsh et al. (2018).

If you were alive in the late 1970s, you may be able to recall the National Geographic special TV show showing scientists on the submersible "Alvin" in the first human encounter in 1977 with a living hydrothermal vent community.[18] I was an undergraduate student at the University of Washington Department of Oceanography at the time, and we students were absolutely enthralled with the story: images of giant clams and tube worms with hemoglobin in their bodies, bleeding red in the claws of Alvin's remote sampling device, and hot fluid venting, turning black in the void of ocean space above. This was the stuff that made a new generation of ocean explorers.

Deep-sea vent communities comprise entirely separate ecosystems, decoupled from solar-powered life on the Earth's surface, having evolved to utilize organic matter synthesized by hydrogen sulfide-reducing bacteria. Over 300 endemic species have been found near the vents, including corals, clams, shrimps, crabs, and the now famous giant, red-tipped tube worms, 2.5 m (8 feet) tall creatures that flourish in waters close to the hot springs.

The communities that inhabit hydrothermal vents exhibit high degrees of endemism (species found only in that particular habitat) and diversity. The average biomass (total weight of all living organisms) associated with vents is an order of magnitude at least larger than that associated with the surrounding abyssal environment. Many animals host symbiotic, chemosynthetic bacteria that convert sulfides into organic matter. Other chemosynthetic organisms make filamentous mats and biofilms that provide food for grazers and deposit feeders.

At the apex of the food web are predators and scavengers such as spider crabs that moved in from the surrounding abyssal areas. Some species of predators and scavengers have evolved to become vent specialists and are found only on vents. The study of vent biology is still in its infancy – only recently it was discovered that some species of skates (related to sharks and rays) use the vents to incubate their eggs.[19] Warm water accelerates egg development. There is still much to learn about the inhabitants of hydrothermal vents.

Apart from providing a totally independent ecosystem from the solar-powered biosphere, hydrothermal vents produce another by-product that is of economic interest to humans: seafloor massive sulfide (SMS) gold and copper deposits. These deposits are formed as the mineral-rich fluid exits the vent and mixes with the ambient seawater and solids are precipitated. The solids appear as white or black "smoke" in underwater video images but are actually mineral-rich particles that have precipitated from the venting fluid. Black smokers are hotter and contain sulfides of iron, copper, and zinc. Minerals in white smoker vents include barium, calcium, and silicon and are emitted from lower-temperature plumes located a greater distance from their heat source.

The precipitates may adhere together to make cylindrical "chimneys" that grow vertically upward from the vent on the seabed. Such chimneys can grow to heights

---

[18] Corliss et al. (1979).

[19] Salinas-de-León et al. (2018).

of over 40 m (120 feet) before falling over and adding their mass to the growing SMS deposits. Mounds of precipitated pyrite-chalcopyrite several meters high cover the seafloor around the vents. Hydrothermal vents and chimneys have now been found in every ocean along all the Earth's spreading ridges. Vent chimneys are usually less than 20 m (60 feet) tall, but one over 45 m (140 feet) in height, named *Godzilla*, was discovered on the Juan de Fuca Ridge, offshore from the US State of Washington. Another group of five giant chimneys, discovered in 2008 on the mid-Atlantic ridge between Greenland and Norway, was named *Loki's Castle* by scientists from the University of Bergen.

The mid-ocean ridge may be a vast, global feature, but the hydrothermal vent communities that evolved to harness the energy of the geothermal heat flux from the Earth are tiny in comparison with the solar-powered biosphere. The total area of rift valleys is estimated to be 710,000 square kilometers (277,000 square miles), but hydrothermal vents are rare, spaced up to 20 km (12 miles) apart along the most active, Pacific ridges, and up to 300 km (180 miles) apart along slower, Atlantic ridges. In other words, there may be as many as 12,000 vent sites in the Pacific, Indian, and Southern Oceans, while the slow-spreading mid-Atlantic ridge supports only about 40 or so hydrothermal vents along its entire length. These are only rough estimates, and the exact numbers are unknown, but the number of vents is finite, and their area is tiny compared with the size of the ocean.

The mineral composition of seafloor massive sulfide gold and copper deposits has attracted the attention of potential deep-sea mining companies, such as Nautilus Minerals, Inc., who have leases to explore prospective areas of the Bismarck Sea in Papua New Guinea. Added to this are mining interests for manganese nodules deposited in the deepest ocean basins, typically below a depth of 4000 m. Manganese nodule deposits often lie outside the jurisdiction of any country (i.e., beyond 200 nautical miles from land) and are managed by the United Nations International Seabed Authority (ISA). Interest in mining the ocean floor is growing. In 2001 there were just six deep-sea mineral exploration contracts, and by the end of 2017, there were 27. The largest lease area awarded by the ISA covers an area of over 80,000 square kilometers (24,000 square miles), about the size of South Carolina.

Commercial deep-sea mining has not yet begun anywhere in the ocean, but testing of equipment to be used in the extraction of the ore has been carried out. It includes giant, three-story tall robot excavators that would churn through the ocean floor, pulverizing the rock and pumping it up to a mother ship floating 2000 m (6000 feet) above on the ocean surface. Deep-sea mining operations would create plumes of suspended sediment that could potentially drift in bottom currents for hundreds of kilometers across the ocean floor, smothering animals along the way. Scientists who have studied deep ocean animals have issued a warning that such mining operations would inevitably lead to loss of biodiversity including the possible extinction of species not yet known to science.[20]

---

[20] Van Dover et al. (2017).

The possibility of deep-sea mining is one of many activities that humans have contemplated using the oceans for that is morally questionable (disposal of nuclear waste in deep ocean trenches, scuttling ships and decommissioned oil rigs, and dumping garbage at sea are among other ideas that have been floated over the years). Yes, it is technologically possible to mine the ocean floor, but should we do it? Is it worth the risk to the unseen and unknown animal life and the possible extinction of species? The resources are nonrenewable, meaning that once extracted they are gone forever. What is the risk to valuable (renewable) fisheries? What about our obligations to future generations? Deep-sea mining falls into the category of questionable human uses of the oceans that will no doubt be debated over years to come.

Chapter 10
An Ocean of Mysteries

"We batter this planet as if we had someplace else to go."
Ann Druyan
Introduction to The Varieties of Scientific Experience by Carl
Sagan, 2006

"Conserve and sustainably use the oceans, seas and marine
resources for sustainable development."
United Nations Sustainable Development Goal #14, UN 2030
Agenda

Abstract This last chapter will review the strange and fickle (geologic) times we live in and the future of our oceans. In the next century, climate change will cause the demise of coral reefs, warmer and more acidic oceans, the poleward migration of species, melting of ice sheets, and thus at least 3 m (10 feet) of sea level rise. There is no such thing as "sustainable growth" in a finite ocean, and so we simply must choose which future we want to have for our civilization and our ocean. Further into the future, we can predict a new ice age beginning after around 10,000 years. Even further into the future, plate tectonics predicts a supercontinent "Pangea Proxima" will be formed in a few hundreds of million years. The oceans will evaporate when the sun becomes a "red giant" in around 1 billion years. The oceans remain largely unexplored, and many mysteries remain for future generations to solve.

Keywords Climate change · Ocean warming · Arctic sea ice · Ocean acidification · Sea level rise · Coral bleaching · Greenland ice sheet · West Antarctic ice sheet · Ecological tipping point · Planetary boundaries · Future ocean · Pangea Proxima

In this book we have considered the main events in the history of the ocean, from its birth shortly after the Earth formed in the Hadean, up to the present time. We started our discussion of the history of the ocean on the 1.1-km-long geological time walk outside the Geoscience Australia building in Canberra and the early bombardment of Earth, when the oceans boiled. Now that we have reviewed many other events and the features of the ocean, let's continue our stroll along the time walk. At the very start of the walk, the Earth formed from a swirling mass of gas, dust, meteors, etc.

We don't get very far, only 22 m along the walk, before the young Earth collided with a Mars-sized planet called Theia, giving birth to the moon. The molten surface of the Earth gradually cooled, and an atmosphere was formed of hydrogen, methane, and water vapor. At around 73 m, the surface had cooled enough for falling rain to accumulate, and the oceans were formed. The bombardment by meteors and comets caused the oceans to boil. By 80 m along the walk, the ocean covered almost the entire surface of the Earth apart from some small, stringy continents and scattered volcanic islands awash with large tides (due to the close orbit of the moon, although we are not sure exactly how the Earth-moon system evolved).

The intense meteor bombardment ended at about 145 m, and life, including photosynthesizing cyanobacteria, arose in the ocean immediately after this – exactly how and when this occurred is matter of debate. Ocean life exhaled oxygen which combined with the masses of iron dissolved in seawater. By 168 m the first layers of iron oxide and silica were deposited on the seafloor that became the earliest beds of the banded iron formation. Stromatolites formed in protected shallow waters after around 206 m. Already, life was leaving its mark in the ocean.

Plate tectonics commenced (we are not sure how or when), and by 366 m the first supercontinent, Ur, appeared. At 440 m the supercontinent Kenorland appeared, with its superocean, Lerova. Oxygen levels built up in the ocean after all the free iron was converted to iron oxide. Then oxygen escaped gradually into the atmosphere, and it began to oxidize atmospheric methane, changing the color of the sky from orange to blue. The methane, a strong greenhouse gas, had acted to keep the climate warm. Its removal resulted in global cooling, and at 516 m, the Earth became ice-covered, entering its first "snowball" phase with glaciers located at the equator. The so-called Huronian glaciation lasted for 300 million years until close to the halfway point of our walk at 550 m.

At 587 m the supercontinent Nuna appeared, and this marked the start of the "boring billion" which takes us round to the homestretch at 904 m. The boring billion saw the first multicellular life, along with the first sexual reproduction at about 756 m, followed by the appearance of the supercontinent Rodinia at 855 m. The most exciting life form then in existence, the stromatolites, reached their peak at around 802 m – all life that existed on Earth was oceanic life.

Life first emerged in the ocean 3.8 billion years ago, and 2.6 billion years later, life had only evolved to this low level of complexity. For the whole of that 2.6 billion years, the oceans persisted, providing the liquid water environment needed for life to exist. Discussions about the likelihood of life having evolved on other planets often mention the need for the existence of liquid water. It should not be overlooked that there is also the need for that liquid water environment to exist for several billions of years to allow the time needed for multicellular life to evolve. One billion years or even 2 billion was not enough time for complex life to evolve, at least on our Earth.

Sometime before 940 m, the Earth cooled off again for reasons as yet unknown, and a second global glaciation (snowball Earth) prevailed for several millions of years. Seasonal sea ice and glaciers formed at tropical latitudes. Life on Earth was still fairly simple stuff, mostly just bacteria and algae. And then, suddenly, at 968 m life went crazy in the Cambrian explosion; by 974 m all the major phyla of animals had appeared. By 990 m the first algae were growing in shallow lakes – finally, the land was colonized by the ocean.

At 1018 m the supercontinent Pangea appeared, surrounded by the Panthalassa Ocean. The ocean contained its first coral reefs, fish, and sharks. Atmospheric oxygen reached an all-time peak value of 35%, and on land, giant insects, flowering plants, and tree ferns evolved, marking the Carboniferous Period. By 1039 m Pangea had split into two continents, Gondwana and Laurasia, with the Tethys Ocean located between them. Dinosaurs evolved and many species adapted to life in the ocean.

A large meteor collided with the Earth at 1084 m, landing on the Yucatán Peninsula in what is now Mexico, causing the extinction of the dinosaurs, including all of the large, swimming dinosaurs. Most of the species of animals that exist today have evolved since then, in the last 16 m of the walk. Plate tectonics split Gondwana and Laurasia into the various continents forming the Atlantic, Arctic, Indian, and Pacific Oceans. The once great Tethys Ocean was squeezed out of existence at 0.5 m before the end of the walk, leaving the Mediterranean Sea as a remnant.

Since the average age of the ocean floor is only 80 million years, we know that most of what we see today in the modern ocean basins evolved in the last 19 m of the walk. The ocean basins and everything that they contain, all the mid-ocean ridges, seamounts, plateaus, and canyons, are relatively recent features, all less than 180 million years in age. Antarctica became isolated at the South Pole, cooled by the formation of the circumpolar current, and by 40 million years ago (the last 9 m of the walk), it was covered in a permanent ice sheet.

It is not before 13.4 cm from the end of the walk that Earth had finally entered its modern climatic regime – a partial snowball Earth scenario. For reasons we don't completely understand, ice sheets started to grow in the northern hemisphere, and ice ages started occurring periodically every 100,000 years as was explained by Milanković. Ice sheets covered Europe and North America before melting away during interglacial warm periods. The modern Earth, which we are most familiar with, evolved only in the last few centimeters of the walk. Humans evolved in the last 2.4 cm, and all of recorded human history has taken place during the last 1.1 mm of the walk.

Many lessons can be taken from this brief history of the oceans. An obvious message is that there are still many unanswered questions about how the Earth, oceans, and our climate system have evolved. I would emphasize three points. First is that biologically produced oxygen has transformed the ocean, atmosphere, and land surface of the Earth. Once dissolved in the ocean, oxygen combined with dissolved iron, thus removing it from the water column. Next, oxygen changed the chemistry of the atmosphere, transforming its color from orange to blue. Oxidation of minerals on the surface commenced the process of "weathering," which in turn transformed the chemistry of the oceans. Once there was oxygen in the atmosphere and dissolved in the ocean, life gradually evolved. It is clear that there would not have been any Cambrian explosion of life if atmospheric oxygen had not first been available and abundant.

The second lesson is the length of time that this took to occur. Life as we know it simply did not exist for nearly 90% of the Earth's history (until after 974 m into the time walk). It took 4 billion years of the Earth's 4.5-billion-year history before the atmosphere was fit for humans to breathe. Diverse life as we know it, especially life on land, is a very recent occurrence. The ocean, on the other hand, has hosted life, albeit in simple forms, for billions of years. Ocean life then colonized the land.

The third lesson is how very finely tuned, precariously balanced, and poorly understood the Earth's climate system is. The factors involved are numerous and interlinked: the albedo effect, the amount and composition of greenhouse gases in the atmosphere, the rate of primary production and oxygen generation, the configuration of continents, the pattern of ocean surface currents and deep ocean circulation, cloud cover and water vapor in the atmosphere, and the Milanković cycles. The geologic record shows that different (unknown) combinations of these factors (plus other factors, some yet to be discovered) have locked the Earth into extreme hothouse and icehouse conditions that have persisted for 10's to 100's of millions of years. The oceans are highly sensitive to such different climate regimes and have responded by becoming anoxic or ice-covered.

We happen to be alive at a special time in Earth history with a continent centered over one pole and a (nearly) landlocked ocean centered over the other. This configuration has made the Earth particularly sensitive to small changes in Milanković orbital parameters; a small change of only a few degrees Celsius heating or cooling has triggered an alternation between an ice age and warm interglacial period every 100,000 years. Against this backdrop, how will the oceans respond to a global temperature increase of 2 °C (i.e., the limit agreed by countries at the 2016 Paris climate negotiations)?

***

What will the ocean be like by the end of the twenty-first century? Will we be able to curb overfishing, pollution, and climate change? Will we allow deep-sea mining to occur? Will we set aside wilderness areas large enough to conserve the oceans' biodiversity?

Beyond human timescales, what will happen to the oceans in deep time? What will be the consequences of glacial cycles, plate tectonics, the evolution of life, and the death of our sun? These are big and difficult questions, but we can make some predictions based on what we already know. Let's start with the likely effects of climate change.

Humans have released about 600 billion tonnes of carbon into the atmosphere since the beginning of the Industrial Revolution, and atmospheric CO_2 concentrations are now increasing at a rate of three parts per million (ppm) per year. During the ice age, 20,000 years ago, CO_2 concentrations were around 180 ppm, rising to around 280 ppm (the natural background level) when the ice age ended around 10,000 years ago. In other words, the change from ice age to warm interglacial was associated with a rise in CO_2 concentration from 180 to 280 ppm, a change of 100 ppm. But it took 10,000 years for this change to occur, never faster than 0.15 ppm of CO_2 per year as far as we know from ice core records collected in Antarctica and Greenland.

At present the CO_2 concentration in the atmosphere is over 400 ppm. In other words, over the last 150 years, humans have raised the CO_2 concentration in the atmosphere by around 120 ppm. Humans are raising CO_2 concentrations 20 times faster than the fastest natural process that we have ever measured. This rate of change has probably never occurred before in Earth's history (and certainly not in the last 2 million years). Therefore, as far as the current rate of change is concerned (3 ppm of CO_2 per year), the geological record cannot help us to predict what will happen to the oceans. We are in uncharted territory.

Throughout this book I have mentioned observations made by oceanographers and climate experts that indicate how anthropogenic greenhouse gas emissions have already changed our climate, the ocean, and ocean life. Here is an abbreviated list:

1. The ocean is getting warmer. The western boundary currents (like the Gulf Stream) are warming up twice as fast as the rest of the ocean. The surface ocean (upper 700 m) has warmed by 0.168 °C (0.302 °F) in the last 50 years based on oceanographic measurements.
2. Arctic sea ice cover is shrinking. Satellite data show that the area of Arctic sea ice at the end of summer has reduced from 7.5 million square kilometers in 1979 to around 4.0 million square kilometers at present, a reduction of 40%.
3. There is evidence for a global trend of increasing wind speed and increased ocean wave height in recent decades.
4. The ocean has become 30% more acidic in the last 50 years due to the increase of CO_2 in the atmosphere that has dissolved into the ocean.
5. The Greenland and Antarctic ice sheets are melting at an increasing rate; Greenland is melting the fastest and is currently losing around 236 cubic kilometers of ice every year.
6. Sea level has risen about 19 cm (10 inches) in the last 100 years, and the rate is getting faster; it was rising about 1.7 mm/year at the end of the twentieth century and is now rising at around 3.4 mm/year.
7. Coral reefs have passed a tipping point whereby bleaching of corals, due to overheating of surface waters, is occurring at a frequency that exceeds the time needed for reefs to recover.

There are many other observations that could be listed here that demonstrate the impacts that humans have already had on the oceans and Earth's climate. But even this short list illustrates one key point: there is no such thing as a climate change "debate" in the scientific literature. Climate change is an established fact. It is not something that will happen in 2050 or 2100; it is happening right now. The only debate about climate change is focused on the details: topics such as spatial variations of ocean warming, the relative importance of surface ocean freshening in slowing the global ocean conveyor, the confidence limits on the climate models, etc. There is no doubt that climate change is taking place and that humans are the reason why.

What does the future climate hold for the ocean?

What we know for certain is that the oceans will become warmer during the next century. Because of the lag effect discussed in Chap. 8, it will take another 25–50 years for the oceans to begin to respond to the warming inherent in the current level of carbon dioxide in the atmosphere (over 400 parts per million, or ppm) and at least 100 years to achieve a 90% response. We are now committed to this warming, and there is nothing we can do that will stop it from occurring unless we are able to reduce the level of carbon dioxide in the atmosphere.

The immediate effects of this warming will be most notable in the polar seas. Sea ice in the Arctic Ocean is in retreat. It covered 40% less area in 2018 than it did in 1979. By the year 2030, it is expected that Arctic sea ice will completely melt away by summer's end (there will still be winter sea ice every year). Species of fish, birds, and mollusks, formerly found only in the North Pacific and North Atlantic, are migrating into the Arctic to take advantage of the seasonally open water and the new climate. Some Arctic species have no place to retreat and their range is decreasing. Extinction of some species is possible.

The warming of the oceans is causing a poleward migration of species along the coasts of all continents. For species that have adapted to life in coastal and shelf habitats in the northern hemisphere, the poleward migration ends in the Arctic Ocean. Where continents terminate in the southern hemisphere, coastal and shelf species endemic to the southern tips of Africa, Australia (Tasmania), and South America (Tierra del Fuego) will run out of space; at the continent's end, there is no coastal habitat further south, and these species will have no place left to go. This is equivalent to alpine species moving up in elevation to seek suitable (cooler) habitats until they reach the mountain summit.

Sea level is likely to rise by 30 cm (1 foot) by 2050 relative to its position in 1950 and by more than 1 m (3 feet) within the next century; most likely it will be closer to 2 m (6 feet) higher by the end of the twenty-first century[1] because of the partial melting of the Greenland and Antarctic ice sheets. The total amount of sea level rise is from a combination of melting ice plus warming of the ocean; warming will cause the ocean to expand, causing around 60 cm (2 feet) of sea level rise by the end of the century. It's the amount of melting ice that we are less certain of.

The melting of Greenland is likely to continue steadily over the next several centuries causing global sea level to rise by 6 m (20 feet). It may take 1000 years to completely melt away. As far as short-term (next 100 years) sea level rise is concerned, the unknown factor is the stability of the West Antarctic ice sheet. This contains an equivalent of around 4 m (12 feet) of sea level, and it seems unlikely that it will survive intact much beyond the end of the twenty-first century (or even that long) with current rates of carbon emissions and attendant warming.

The collapse of ice sheets is not always a gradual, incremental process. It may instead be very rapid, resulting in the catastrophic flooding of coasts. Glaciologist

[1] Hansen et al. (2016).

Christina Hulbe points out that the existing climate models agree on one thing: "When driven with environmental variables from mid-range and high-end global warming scenarios, all predict collapse of the (West Antarctic ice sheet)."[2] The models are less clear on the timing of collapse and also hold out hope that collapse can be avoided if greenhouse gas emissions are reduced.

The meltwater flowing off Greenland has already started to cool and freshen the North Atlantic surface waters south of Greenland. Wally Broecker concluded that, although the global conveyor will slow down in a warmer world, it is unlikely to stop altogether.[3] The slowdown will be caused by increased rainfall, as heating will drive a faster hydrological cycle, with added meltwater from the Greenland ice sheet. The effects of any slowdown in the global conveyor are unknown, but there will be consequences.

For example, a slowing of the global conveyor implies that the lungs of the ocean will also slow down. Less oxygen will be breathed in, and the ventilation of the deep sea will be more sluggish. Presumably this means oxygen dead zones will expand in the North Pacific, Gulf of Mexico, and elsewhere. It also means that the rate of carbon dioxide uptake by the polar oceans will be slower, reducing the amount taken from the atmosphere. This will only add to the carbon dioxide left in the atmosphere, a positive feedback for global climate change.

A slowing of the global conveyor will no doubt have numerous other unforeseen consequences. For example, if deep ocean bottom currents were to slow by 50%, this would imply a 50% reduction in food supply to filter-feeding benthic animals that rely on food being carried to their waiting tentacles by the passing current. It will mean a 50% reduction in the distance larvae will be dispersed by the deep ocean currents. Will the affected ecosystems be able to adapt and survive?

One final climate-related issue, in addition to oceans becoming warmer, sea level rising, and slowing of the global conveyor, is the acidification of the ocean (the "evil twin" of global warming) caused by humans raising atmospheric carbon dioxide levels. The likely effects that a lower ocean pH will have on species are a relatively new field of research, which has become the focus of a growing scientific effort once the enormity of the threat was realized in the early 2000s. About half of the research has focused on coral reefs and the rest on other species. The picture is gradually becoming clearer and the news is not good.

Coral reefs that have managed to survive repeated bleaching from a warmer ocean will gradually become further weakened by pH stress. Acidification will impact upon all species that build their bodies from calcium carbonate including planktonic coccolithophores, pteropods and other mollusks, echinoderms, and coralline algae. The potential impacts on communities and ecosystems are not well understood.

The science has been done. We know everything we need to know about the degradation of the oceans to accept responsibility for how things are. It is now a question of choice – what will we humans do next? We have passed 400 ppt of

---

[2] Hulbe (2017).
[3] Broecker (2005).

carbon dioxide in the atmosphere. What will it be by the year 2050? Or by the year 2100? Or by the year 3000? It makes no sense to plan only for the next few decades when we all desire to see our civilization continue indefinitely. Will humanity take action to reduce greenhouse gas emissions before we pass an ecological tipping point that compromises our civilization? Answering this question has been described by Al Gore and other world leaders as the greatest moral challenge of our time. It probably is.

But climate change is only one challenge humans face in managing our use of the oceans. There are many others. In earlier chapters I have mentioned pollution from oil spills, noise from ship propellers and seismic surveys, the scourge of plastic in the ocean, overfishing, sand mining, and the looming threat of deep-sea mining. These issues are no less important than climate change, and they are recognized as such by some parts of society. The problem is that marine life is often impacted by several or all of these human activities at the same time – the effects are cumulative.

There are United Nations agreements and national policies that are in place to address some of these problems. However, in many cases they do not go far enough, and passing resolutions is not the same thing as providing solutions. Once again, it seems the ocean's opaqueness is working against it. The damage we cause cannot easily be seen, and so society is less inclined, or at least not sufficiently motivated, to take prompt and decisive action.

The future, then, is that one by one, for each issue, we will come up against the ocean's natural limits. Unless we choose to curb our impacts and set aside areas for conservation and restoration of ecosystems, we risk the permanent loss of species and the benefits that the oceans provide to humans. We appear to have already passed the limits for the bleaching of coral reefs, year-round Arctic sea ice, and oxygen levels in some locations of the oceans. We have already caused the extinction of some ocean species (many are likely unknown to us), and others are under dire threat. We must accept that the ocean is finite and that it has limits.

***

Ann Druyan is right.

We treat the Earth (and ocean) as if we had a spare one, just in case things go wrong. As if we had a "Plan B." As if, were it to turn out that the planet we have now can't cope with the way we mismanage it, we have another option.[4]

But, of course, we do not. There is only one Earth and it has one ocean. This is the only place in the solar system where we can survive.

The ocean is also finite. There is no such thing as "sustainable blue growth" in a finite ocean. We must acknowledge that there are limits and learn to behave accordingly. Humans already use 75% of the arable land on Earth and consume 25% of the Earth's terrestrial primary production,[5] as well as between 24% and 35% of coastal and shelf marine primary production. We consume 25% of the available freshwater.

---

[4] Borja and Elliott (2018).
[5] Krausmann et al. (2013).

We humans cannot indefinitely expand the proportion of Earth's resources we take for ourselves if other species are to survive. There are limits to how many fish we can take, how much waste we can dispose of, and how much excess CO_2 gas the ocean can absorb before we reach a tipping point. These limits underpin the concept of "planetary boundaries,"[6] a very useful way to think about life on our Earth and the systems needed to sustain it. This concept is also associated with the notion of "sustainable use" as enshrined in the United Nations 2030 Agenda including Goal 14 for sustainable use of the oceans.

The oceans today cover 70% of the surface of Earth, and they have an average depth of about 4000 m (12,000 feet), which means they contain approximately 1.3 billion cubic kilometers of water. When you consider that there are now about 7 billion people living on the Earth, there is about one-fifth of a cubic kilometer for each of us. One-fifth of a cubic kilometer seems like a lot of water: think of a box that is 1 km wide, 1 km long, and 200 m tall. That box contains 200 million liters of water!

But then consider how utterly dependent upon the oceans you are; that small one-fifth of a cubic kilometer produces half of the oxygen you breath (every second breath you take) and all of the seafood you eat, and it is the ultimate source of all the freshwater you will drink in your lifetime. It will also receive your share of the sewage, garbage, plastic, spilled oil, and industrial waste that is dumped into the ocean. We each need to take very good care of our fifth of a cubic kilometer of ocean!

Our review of the history of the ocean illustrates how short the time has been that we could actually survive on Earth – out of 4.5 billion years of Earth's history, it has only been possible for us to breathe the atmosphere for the last 400 million years or so. We are lucky to be alive during a geologic epoch when the Earth is not in a full icehouse state, with sea ice covering most of the ocean. But we're not living in a total hothouse either. Having an ice sheet conveniently located on Antarctica has lowered sea level by around 70 m, exposing the flat coastal plains and arable land where most of us live and where we grow most of our food. Today, we are lucky to have 30% of the Earth as dry land, compared with only 18% that the dinosaurs had when there was no continental ice sheet anywhere on Earth and the low-lying coastal plains were submerged beneath shallow seas, flooded by an overflowing ocean.

We have learned how fickle the Earth's climate can be. We know that we are dealing with a highly sensitive system that has changed in the past in dramatic and dangerous ways that have caused extinctions and wreaked general havoc on the biosphere. Drastic changes in climate have occurred in response to seemingly small changes in the amount of meltwater discharged from melting glaciers or in the concentration of greenhouse gases. It is a mistake to think that raising the carbon dioxide level by one more tiny part per million will yield a correspondingly tiny increase in global temperature. That is not how Earth systems respond – rather, once a tipping point is passed, the response is more likely to be nonlinear, potentially catastrophic, and irreversible over timescales of human life spans.

I would conclude this discussion on a positive note, because I am fundamentally optimistic about humanity's future. There is an old saying that the Stone Age did not

---

[6] Steffen et al. (2015).

end because we ran out of stones. Humanity simply found a better way. I believe that the "Age of Oil" will also end, not because we will run out of oil but because we will find a better way. Human ingenuity will find solutions to halt the loss of species and curb pollution, including greenhouse gasses like CO_2. I only hope we find the solutions we need and have the courage to put them into effect, before too much damage is done to our fragile ocean world.

***

What will the ocean look like 10,000 or 100,000 years in the future? To answer this question, we turn to the governing processes discovered by Milanković, which are dominant at this timescale.

As discussed in Chap. 7, the Earth's climate is presently extremely sensitive to small variations in solar heating. According to Milanković theory, when the Earth is further from the sun when it is summer in the northern hemisphere, the snow that fell the previous winter does not melt and glaciers advance. Over time the world descends into an ice age. Modelling of the Earth's orbital patterns indicates that within the next 12,000 years, conditions will favor the Earth entering another ice age.

What we have learned from the study of deep ocean sediments and ice cores from Antarctica and Greenland is that the glacial cycle is not a simple on/off switch. The amount of time that it takes for the Earth and ocean to exit from the most extreme ice age condition is far less than the amount of time it takes for the Earth and ocean to enter a new ice age. The descent into an ice age is gradual and complex with many changes with glaciers waxing and waning. Changes in the volume of ice on land mean that global sea level also falls and rises again during a gradual descent into a glacial phase. The growth of ice sheets on land causes sea level to fall, and melting ice sheets on land cause sea level to rise (Fig. 10.1).

Over the last glacial cycle, the descent into an ice age took around 80,000 years, starting from the last interglacial period about 125,000 years ago up to the last glacial maximum about 18,000 years ago. In contrast, the melting of the northern hemisphere ice sheets took a mere 10,000 years, bringing an end to the ice age. We are presently in an interglacial period which is comparable to the situation around 120,000 years ago. To put it simply, the pattern for the last 1 million years is that ice ages come creeping up slowly and end with a bang!

What does this tell us about the possible future of the ocean over the next 10,000–100,000 years? Over the next 1000 years, humans will have to stop burning fossil fuels at the rate we are burning them today. We will do this for either one of two reasons: (1) because we take wise decisions to curb our burning of fossil fuels or (2) because they will run out. There is only so much oil and coal in the ground, and once it's gone, we will have no choice but to change to renewable energy sources.

The warming resulting from the increased CO_2 will cause sea level to quickly rise at least 6–10 m above its present level because the Greenland and West Antarctic ice sheets will melt. I mentioned above that some melting of the Greenland and West Antarctic ice sheets is now inevitable by the end of this century and into the twenty-second century. It will take several thousands of years for the ocean and atmosphere to equilibrate to CO_2 levels once humans curb burning of fossil fuels.

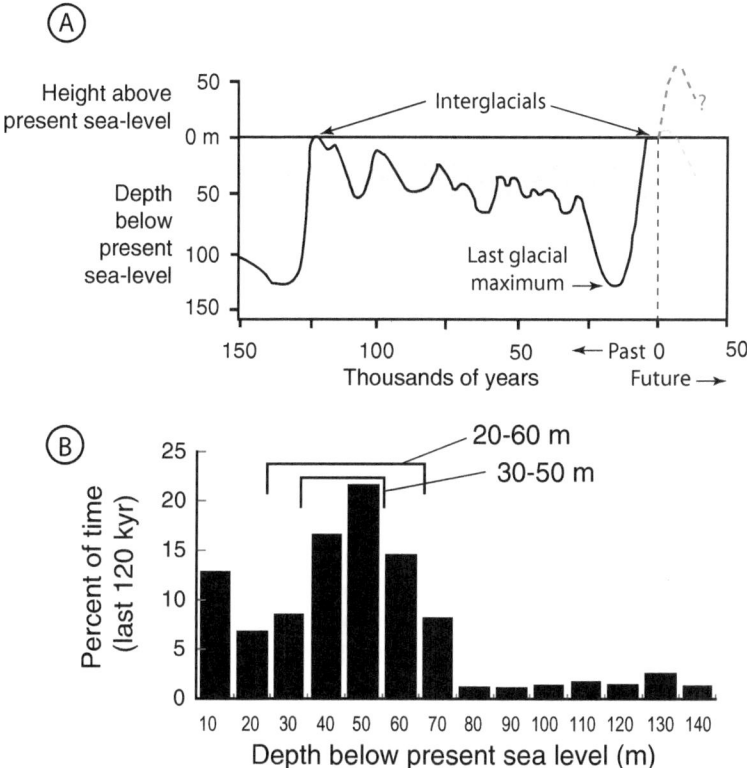

Fig. 10.1 Global sea level curve for the last 150,000 years. (From Chappell and Shackleton 1986). The blue line represents the likely future sea level curve with melting only the Greenland and West Antarctic ice sheets, rising sea level by around 6–9 m, depending on the amount of CO_2 humans release into the atmosphere. The red line represents the worst-case scenario where the East Antarctic ice sheet is melted which will happen if humans do not curb CO_2 emissions. (**b**) Histogram showing percentage of time that sea level has been within 10 m depth bands (i.e., 0–10 m, 10–20 m, etc.) over that past 120,000 years (equal to one full glacial cycle), based on the curve shown above. The data show that sea level was within the 30–50 m depth range for approximately 38% of the time (46,400 years) over the past 120,000 years and within the 20–60 m depth range for approximately 60% of the time (74,500 years, the area shown in gray shading in **a**). By comparison, sea level has been in the 0–10 m range for only 12.8% of the time (15,500 years)

Over the next few thousand years, natural processes will restore atmospheric CO_2 to pre-industrial levels, and the anthropogenic greenhouse effect will end.

What happens if we don't curb our CO_2 emissions? What if we burn all the fossil fuels in the ground? Then we can expect CO_2 levels to exceed 1000 ppm. If humans are unable to curb CO_2 emissions needed to slow the acceleration of climate change, a substantial part of the East Antarctic ice sheet could melt raising sea level 70 m higher than at present. Judging from the amount of time needed to melt the northern hemisphere ice sheets at the end of the last ice age (which took about 10,000 years), we can expect that melting the East Antarctic ice sheet could happen in less than 10,000 years.

This worst-case scenario puts us into uncharted territory because we have had an ice sheet in Antarctic for over 30 million years. The last time we had no polar ice caps was back in the Eocene when CO_2 reached around 700–900 ppm. The Earth would be a very different place with temperatures probably unbearable on the equator and subtropical at the poles. A 70 m rise of global sea level (60 m from melting East Antarctica added to the 10 m from melting the Greenland and West Antarctic ice sheets) would flood our best farmland and reduce land area by 30%.

About 12,000 years in the future, the natural Milanković climate forcing will take over, and Earth will begin its descent into an ice age. Sea level is likely to fall by at least 20 m and probably more. We know this because sea level has been 20–60 m below its present level for at least 60% of the time during the last glacial cycle (the last 120,000 years) and it is likely that this pattern will continue. The next glacial maximum, with ice sheets covering the northern hemisphere and sea level around 120 m below its present position, will not occur for another 80,000 years or so.

This is the prognosis if humans leave the climate in the hands of natural forcing and do not intervene to control it. Could humans halt the coming of an ice age? Climate control might become feasible in the future by artificially maintaining elevated greenhouse gas levels in the atmosphere or by using some combination of new, yet-to-be-invented technologies. But if we do not interfere with the natural Milanković climate forcing, the Earth will descend into an ice age.

Standing on a shoreline 10,000 years into the future, what are we likely to behold? Sea level would finally be falling because the ice sheets are beginning to regrow. How strange it is to see our twenty-first-century harbors and coastal cities gradually revealed by the falling tide. As for life in the sea, the species we can see are only a subset of those alive today because many have gone extinct. Acidification of the oceans, loss of habitats, overfishing, pollution, and other pressures will overwhelm and kill many species. But many species will have survived, and some will have prospered due to wise use of the ocean's fisheries by future humans. The shallow shelf seas are the most productive part of the ocean, and their area will have greatly expanded at the expense of low-lying coastal land because sea level is much higher.

Standing on a shoreline 10,000 years into the future, the view we behold will depend a great deal upon the decisions we take today.

***

What future oceans will there be in deep time? What will the oceans and configuration of continents look like 1 million years or 100 million years in the future? To answer these questions, we turn to plate tectonics and the evolution of our sun which are the dominant governing processes at these timescales.

Since we know where the mid-ocean spreading ridges are located and approximately how fast ocean crust is forming along them, we can use the present vectors of the continents to predict where they will be in the geologic future. This is exactly what geologist Christopher Scotese has done to create an animation of how the future configuration of oceans and continents might appear.[7]

This projection extends 250 million years into the future culminating in the formation of a new supercontinent "Pangea Proxima." In this model Africa collides with Southern Europe, squeezing the Mediterranean out of existence in about 50 million years. Australia collides with Southeast Asia, and the Atlantic and Arctic Oceans continue to expand. By 100 million years in the future, the expansion of the Arctic and Atlantic Oceans has fused the continent of Africa + Eurasia + Australia into a single landmass and moved it into a position that straddles the equator. Meanwhile, Antarctica is finally dislodged from its position on the South Pole and begins to move northward.

Then, at around 150 million years in the future, the scenario calls for the creation of a subduction zone in the western Atlantic, all along the eastern margin of North and South America. This causes a reversal for the Atlantic Ocean which has attained its maximum size and which now begins to rapidly shrink, which, in turn, causes the Pacific to expand.

North America is finally parted from Eurasia along the mysterious plate boundary that extends through Siberia. Rejuvenated seafloor spreading expands the Pacific Ocean which reclaims its dominance as the largest ocean. Antarctica moves north and collides with the giant continent of Africa + Eurasia + Australia. Merging of all the continents finally occurs about 250 million years in the future – there is once again only one, world-covering, "Panthalassa-Proxima" Ocean.

What about future life forms on Earth? It has been posited that we are living now in the midst of the sixth global extinction event, caused by humans. If this is true, then once the extinction event is over, the recovery of life may take many millions of years. For example, it took around 30 million years for life to recover after the end of Permian extinction that occurred 250 million years ago. After the asteroid collision that killed the dinosaurs, it was around 10 million years before life, and ecosystems were fully recovered. It is not clear that the present human-induced extinction event will be as devastating for life as the end-Permian or the end-Cretaceous events, but the geologic record suggests recovery will require millions of years (rather than, say, merely thousands of years).

Extinction events are bad news for the species that dominated the biosphere prior to the extinction event. But in every extinction event in the geologic record, life has afterward recovered, and new groups of species have evolved. As long as Earth retains its ocean and atmosphere, it seems reasonable to predict that life will continue to exist here in some form. But it may not be us!

Finally we come to the ultimate limiting factor for the future existence of the ocean: the sun. Astronomers calculate that our sun is about halfway through its life span, but well before its explosive ending, our star will enter an evolutionary phase that will destroy the Earth. In its later life stages, in around 3.5 billion years from now, our star will expand to become a "red giant," with its surface beyond the

---

[7] Scotese (2014).

present orbit of Mars. Earth will obviously be consumed by the sun well before then. In fact, the ocean probably has less than 1 billion years left to survive.

Over the next 1 billion years, the sun's brightness will increase, heating the Earth. The heat will cause the ocean to evaporate, raising the water vapor content of the atmosphere. Water vapor is a powerful greenhouse gas, and so the heating and evaporation of the oceans will accelerate. Will the ocean boil once again, as it did during the comet bombardment of the young planet? Perhaps.

As the ocean shrinks in size, the continental shelves will be exposed followed gradually by the continental slopes. The ocean basins will gradually become disconnected. The process will depend on the configuration of the continents at the time, which will in turn determine the size and connectedness of the ocean basins. Evaporation will leave the dissolved salts behind, and so the ocean salinity will increase becoming brine. In the final stages, only the deepest ocean trenches will remain flooded with hypersaline water, making long skinny-shaped oceans clinging to the active plate margins of the last continents.

One billion years from now, the ocean will have evaporated entirely, and the surface of the Earth will be devoid of ocean and atmosphere and, therefore, of life.

***

How fast do you think you are moving right now? Probably faster than you think. If you are sitting at home in your chair and you happen to live on the equator, the rotation of the Earth is carrying you around the Earth's circumference once a day at a speed of 460 m/s or roughly 1000 miles/hour. But if your chair is located exactly at the South or North Pole, the rotation of the Earth is simply turning you around in a circle once a day, and your speed is zero. On the other hand, the Earth is orbiting the sun, so, relative to the sun, all of us earthlings are moving through space at a speed of nearly 30 km/s or 67,000 miles/hour.

Furthermore, our sun (along with the Earth and other planets in our solar system) is located about two-thirds the way along a spiral arm (the Orion Arm) that moves around the center of our Milky Way Galaxy. We are spinning around the Galaxy at some 220 km/s (490,000 miles/hour). If the average life span of a human is 70 years, then each of us can expect to travel 485 billion kilometers (about 300 billion miles) in one lifetime. It takes the sun around 226 million years to orbit the Milky Way just once, about 3.2 million human lifetimes. In the entire history of the human species, we have yet to complete even a small part of one orbit of our galaxy. In Earth's 4.5-billion-year history, our sun has completed only 17 orbits of the Milky Way.

An old Breton fishermen's prayer reads: "Oh God thy sea is so great and my boat is so small." But maybe we should remember the prayer like this: "Oh God thy galaxy is so great and my planet is so small." Because in a very real way, we are all living on board the spaceship Earth, travelling around the Milky Way Galaxy. And just as any fisherman depends upon the condition of his boat to survive the vastness of the sea, we are each of us totally dependent upon the condition of our spaceship Earth as we travel through the vastness of space. If we don't keep our spaceship in good shape and functioning properly, it may stop providing the essential things we need to survive on our voyage.

I think that we know all we need to know about how humans have harmed the ocean in recent decades to make informed decisions about what needs to be done to turn things around. The choice is ours. Do we want to maintain life and functioning ecosystems in our oceans for future generations? If we do, then we must drastically reduce our use of fossil fuels. Sustainable *development* means replacing carbon-emitting energy sources with renewable energy sources. It means setting aside marine parks to protect and conserve ocean life and biodiversity. It means that our economic systems must evolve from being growth-based to a circular economic model.

You might ask, if all the science is done, then what it the point of doing any more? Don't we already know all we need to know about the oceans? It is a reasonable question to ask scientists to justify the work they do when it is paid for by taxpayers. The question is, if there is no economic reason to pursue a particular line of science, then why do it?

The answer is, of course, because we have bigger questions to answer than simply "are we doing harm?" Many unanswered questions remain, and the collection of more information is needed to address them. Recall, for example, the search for lost aircraft MH370 (Chap. 4) and the lack of a global map of the seafloor suitable for such search and rescue operations.

Sustainable development means we need more science, not less. Science underpins economic development. Advances in technology are the driving force for economic growth.

The other reason is because we are curious. Because as humans, we need to understand. We need to know. And as happens so often in science, economic benefits follow close behind the discovery of new knowledge and understanding (recall Chap. 4 how scientific curiosity drove Doc Ewing in the development of seismic profiling technology that now underpins the offshore oil and gas industry).

Perhaps the greatest and most wonderful thing about the human condition is our ability to ask questions and seek answers about the oceans, the world, and the universe. The ability to reason is a precious gift, and when it is combined with creative imagination, it is a potent force for solving mysteries. To seek understanding is so much more satisfying than merely accepting an explanation on blind faith. The more that a person understands about the world, the less likely she or he is to accept dogma without question or to believe in anything that cannot be logically explained.

Scientific knowledge is a fundamentally different way of understanding the world than art or religion. If all the world's religious texts, music, and artworks were somehow destroyed, they could never be replaced. Humans would of course create new artworks and invent new religions, but they would not be the same as the ones that were lost.

The same is not true for scientific knowledge.

If all the scientific knowledge we have now was somehow destroyed, future generations would be able, through observation and reasoning, to recreate it. The theories of evolution, plate tectonics, special relativity, and the laws of physics would eventually be rediscovered, working from first principles and applying the scientific method. Science is the only true universal language. The scientific knowledge of sentient alien species living on other planets is exactly the same as our own (albeit further or lesser advanced).

The desire to solve one or other of the ocean's mysteries has inspired generations of adventurers, explorers, scientists, and sailors. They have crossed uncharted seas and suffered great privation, descended in flimsy devices into the inky blackness of the abyss, and spent uncounted weary hours taking measurements and observations of sea life, seawater, and the seafloor. Some have died in the process; others survived to try again and again. Each of them has peeled away some little bit of mystery to reveal new wonders and, more often than not, new mysteries. And so the oceans remain mysterious, holding back secrets for future generations to ponder and investigate.

And in the end, a little mystery is probably a good thing. Because we humans need a little mystery in our lives. It keeps us thinking and wondering about nature and our place in the universe. Our lives are all the richer for the unsolved mysteries that call us away from home and across the sea to explore an unknown place and perhaps to make a small discovery that we might call our own.

Acknowledgments

Scientific innovation and discovery are inevitably incremental, based upon the cumulative work of many individuals who came before. This is particularly true in the ocean sciences where research is conducted mainly by teams; no oceanographer has ever gone to sea without the support of at least a ship, captain, and crew. It has been an honor for me to have sailed on the following research vessels in my career: *RV Onar, RV Venturous, RRS Fredrick Russel, RRS Discovery, RV Rig Seismic, RV Franklin, HMAS Cook, HMAS Flinders, AM Brolga, RV Sea Wanderer, RV Sunbird, RV Southern Surveyor, MV Endeavour Pearl, MV Western Venturer, RSV Aurora Australis, RV Blue Fin, ARSV L M Gould, RV Tangaroa, RV Melville, RV James Kirby, and RV Tully.* Where I have given names of individuals in this book, they are either my co-workers (shipmates) or people who have added an important piece of knowledge to an already partially completed puzzle. For my part, I owe debts to many people from whom I have learned so much. I must thank my wife Ellen, for her unwavering support, the late Prof. Mike Collins who was my PhD supervisor and mentor over many decades, and the many people with whom I have had the pleasure to work with at sea and on land, to study the oceans, and to help explain some small mysteries.

P. T. Harris, *Mysterious Ocean*, https://doi.org/10.1007/978-3-030-15632-9

References

Allison, I. (1999). *RSV Aurora Australis voyage 1-1998-99 report*. Australian Antarctic Division. https://data.aad.gov.au/aadc/reports/get_file.cfm?report_id=4259

Baker, E. K., & Harris, P. T. (1991). Copper, lead and zinc in the sediments of the Fly River Delta and Torres Strait. *Marine Pollution Bulletin, 22*, 614–618.

Baker, E. K., Puglise, K. A., & Harris, P. T. (2016). *Mesophotic coral ecosystems — A lifeboat for coral reefs?* (p. 98). The United Nations Environment Programme and GRID-Arendal, Nairobi and Arendal. https://www.grida.no/publications/88

Barber, D. C., Dyke, A., Hillaire-Marcel, C., Jennings, A. E., Andrews, J. T., Kerwin, M. W., Bilodeau, G., McNeely, R., Southon, J., Morehead, M. D., & Gagnon, J. M. (1999). Forcing of the cold event of 8,200 years ago by catastrophic drainage of Laurentide lakes. *Nature, 400*, 344–348.

Becker, J. J., Sandwell, D. T., Smith, W. H. F., Braud, J., Binder, B., Depner, J., Fabre, D., Factor, J., Ingalls, S., Kim, S. H., Ladner, R., Marks, K., Nelson, S., Pharaoh, A., Trimmer, R., Von Rosenberg, J., Wallace, G., & Weatherall, P. (2009). Global bathymetry and elevation data at 30 arc seconds resolution: SRTM30_PLUS. *Marine Geodesy, 32*(4), 355–371.

Bills, B.G. & Ray, R.D. (1999), Lunar orbital evolution: A synthesis of recent results. *Geophysical Research Letters, 26*(19), 3045–3048, Bibcode:1999GeoRL.26.3045B. https://doi.org/10.1029/1999GL008348.

Blamey, N. J. F., Brand, U., Parnell, J., Spear, N., Lécuyer, C., Benison, K., Meng, F., & Ni, P. (2016). Paradigm shift in determining Neoproterozoic atmospheric oxygen. *Geology, 44*(8), 651.

Borja, A., & Elliott, M. (2018). There is no Planet B: A healthy Earth requires greater parity between space and marine research. *Marine Pollution Bulletin, 130*, 28–30.

Broecker, W. S. (1991). The great ocean conveyor. *Oceanography, 4*, 79–89.

Broecker, W. S. (2003). The oceanic CaCO3 cycle. In H. D. Holland & K. K. Turekian (Eds.), *Treatise on geochemistry* (pp. 529–549). Amsterdam: Elsevier.

Broecker, W. S. (2005). *The role of the ocean in climate yesterday, today and tomorrow*. Palisades: Eldigio Press, Lamont-Doherty Earth Observatory, Columbia University.

Broecker, W. S., Peacock, S. L., Walker, S., Weiss, R., Fahrbach, E., Schroeder, M., Mikolajewicz, U., Heinze, C., Key, R., Peng, T. H., & Rubin, S. (1998). How much deep water is formed in the Southern Ocean? *Journal of Geophysical Research, 103*, 15,833–815,843.

Buffet, J. (1974). A pirate looks at forty. Dunhill Records (music).

Burke, L., Reytar, K., Spalding, M., & Perry, A. (2011). *Reefs at risk revisited* (p. 130). Washington, D.C.: World Resources Institute. http://pdf.wri.org/reefs_at_risk_revisited_executive_summary.pdf

© Springer Nature Switzerland AG 2020
P. T. Harris, *Mysterious Ocean*, https://doi.org/10.1007/978-3-030-15632-9

Cairns, S. D. (2007). Deep-water corals: An overview with special reference to diversity and distribution of deep-water scleractinian corals. *Bulletin of Marine Science, 81*, 311–322.

Carey, R. J., Wysoczanski, R., Wunderman, R., & Jutzeler, M. (2014). Discovery of the largest historic silicic submarine eruption. *Eos, 95*, 157–164.

Carson, R. (1951). *The sea around us* (p. 288). Oxford University Press, New York.

Chappell, J., & Shackleton, N. J. (1986). Oxygen isotopes and sea level. *Nature, 324*, 137–140.

Cooper, A. K., & O'Brien, P. E. (2004). Leg 188 synthesis: Transitions in the glacial history of the Prydz Bay region, East Antarctica, from ODP drilling. In A. K. Cooper, P. E. O'Brien, & C. Richter (Eds.), *Proceedings of the ocean drilling program, scientific results* (pp. 1–42). College Station: Ocean Drilling Program.

Corliss, J. B., Dymond, J., Gordon, L. I., Edmond, J. M., von Herzen, R. P., Ballard, R. D., Green, K., Williams, D., Bainbridge, A., Crane, K., & van Andel, T. H. (1979). Submarine thermal springs on the Galapagos rift. *Science, 203*, 1073–1083.

Dai, A., & Trenberth, K. E. (2002). Estimates of freshwater discharge from continents: Latitudinal and seasonal variations. *Journal of Hydrometeorology, 3*, 660–687.

Davies, P.J. (1974). Subsurface solution unconformities at Heron Island, Great Barrier Reef. *Proceedings of the 2nd International Coral Reef Symposium*, pp. 573–578.

Davies, P. J., Symonds, P. A., Feary, D. A., & Pigram, C. J. (1987). Horizontal plate motion: A key allocyclic factor in the evolution of the great barrier reef. *Science, 238*, 1697–1700.

de Lavergne, C., Madec, G., Capet, X., Maze, G., & Roquet, F. (2016). Getting to the bottom of the ocean. *Nature Geoscience, 9*, 857–858.

Doney, S. C., Fabry, V. J., Feely, R. A., & Kleypas, J. (2009). Ocean acidification: The other CO_2 problem. *Annual Review of Marine Science, 1*, 169–192.

Dysthe, K., Krogstad, H. E., & Müller, P. (2008). Oceanic rogue waves. *Annual Review of Fluid Mechanics, 40*, 287–310.

Edwards C. T., Saltzman M. R., Royer D. L., Fike D. A. (2017). Oxygenation as a driver of the Great Ordovician Biodiversification Event. *Nature Geoscience*. https://doi.org/10.1038/s41561-017-0006-3.

Ekman, V. W. (1905). On the influence of the Earth's rotation on ocean currents. *Arkiv för matematik, astronomi och fysik, 2*, 1–52.

Felt, H. (2012). *Soundings – The story of the remarkable woman who mapped the ocean floor* (p. 119). New York: Henry Holt and Company. [biography of Marie Tharp].

Forsberg, R., Sørensen, L., & Simonsen, S. (2017). Greenland and Antarctica ice sheet mass changes and effects on Global sea level. *Surveys in Geophysics, 38*(1), 89–104.

Frey, F. A., Coffin, M. F., Wallace, P. J., Weis, D., Zhao, X., Wise, S. W., Wähnert, V., Teagle, D. A. H., Saccocia, P. J., Reusch, D. N., Pringle, M. S., Nicolaysen, K. E., Neal, C. R., Müller, R. D., Moore, C. L., Mahoney, J. J., Keszthelyi, L., Inokuchi, H., Duncan, R. A., Delius, H., Damuth, J. E., Damasceno, D., Coxall, H. K., Borre, M. K., Boehm, F., Barling, J., Arndt, N. T., & Antretter, M. (2000). Origin and evolution of a submarine large igneous province: The Kerguelen plateau and Broken Ridge, southern Indian Ocean. *Earth and Planetary Science Letters, 176*(1), 73–89.

Friedrich, O., Erbacher, J., Moriya, K., Wilson, P. A., & Kuhnert, H. (2008). Warm saline intermediate waters in the cretaceous tropical Atlantic Ocean. *Nature Geoscience, 1*, 453–457.

Fujita, K., Park, J., Coakley, B.J., Bourgeois, J., Ponomareva, V., Mackey, K., Miller, E. (2004). *Joint US-Russia workshop on the plate tectonic evolution of northeast Russia – Report from the neotectonics breakout group*. Stanford: Stanford Univeristy School of Earth Sciences, Stanford University.

Galgani, F., Leaute, J. P., Moguedet, P., Souplet, A., Verin, Y., Carpentier, A., Goraguer, H., Latrouite, D., Andral, B., Cadiou, Y., Mahe, J. C., Poulard, J. C., & Nerisson, P. (2000). Litter on the sea floor along European coasts. *Marine Pollution Bulletin, 40*, 516–527.

Galloway, W. E. (1975). Process framework for describing the morphologic and stratigraphic evolution of deltaic depositional systems. In M. L. Broussard (Ed.), *Deltas, models for exploration* (pp. 87–98). Houston: Houston Geological Society.

Geirsson, H., Árnadóttir, T., Hreinsdóttir, S., Decriem, J., LaFemina, P. C., Jónsson, S., Bennett, R. A., Metzger, S., Holland, A., Sturkell, E., Villemin, T., Völksen, C., Sigmundsson, F., Einarsson, P., Roberts, M. J., & Sveinbjörnsson, H. (2010). Overview of results from continuous GPS observations in Iceland from 1995 to 2010. *Jokull, 60*, 3–22.

Hansen, J., Nazarenko, L., Ruedy, R., Sato, M., Willis, J., Del Genio, A., Koch, D., Lacis, A., Lo, K., Menon, S., Novakov, T., Perlwitz, J., Russell, G., Schmidt, G. A., & Tausnev, N. (2004). Earth's energy imbalance: Confirmation and implications. *Science, 308*, 1431–1435.

Hansen, J., Sato, M., Hearty, P., Ruedy, R., Kelley, M., Masson-Delmotte, V., Russell, G., Tselioudis, G., Cao, J., Rignot, E., Velicogna, I., Tormey, B., Donovan, B., Kandiano, E., von Schuckmann, K., Kharecha, P., Legrande, A. N., Bauer, M., & Lo, K. W. (2016). Ice melt, sea level rise and superstorms: Evidence from paleoclimate data, climate modeling, and modern observations that 2 °C global warming could be dangerous. *Atmospheric Chemistry and Phyics, 16*, 3761–3812.

Harris, P. T. (2011). Benthic environments of the Lord Howe rise submarine plateau: Introduction to the special volume. *Deep Sea Research Part II, 58*, 883–888.

Harris, P. T., & Coleman, R. (1998). Estimating global shelf sediment mobility due to swell waves. *Marine Geology, 150*, 171–177.

Harris, P. T., & Collins, M. B. (1988). Estimation of annual bedload flux in a macrotidal estuary, Bristol Channel, U. K. *Marine Geology, 83*, 237–252.

Harris, P. T., & MacMillan-Lawler, M. (2016). In C. W. Finkl & C. Makowski (Eds.), *Global overview of continental shelf geomorphology based on the SRTM30_PLUS 30-arc second database. Seafloor mapping along continental shelves* (pp. 169–190). Cham: Springer International Publishing.

Harris, P. T., & MacMillan-Lawler, M. (2017). Origin and characteristics of ocean basins. In A. Micallef, S. Krastel, & A. Savini (Eds.), *Submarine geomorphology. Springer geology* (pp. 111–134). Cham: Springer.

Harris, P. T., & Whiteway, T. (2011). Global distribution of large submarine canyons: Geomorphic differences between active and passive continental margins. *Marine Geology, 285*, 69–86.

Harris, P. T., Baker, E. K., Cole, A. R., & Short, S. A. (1993). A preliminary study of sedimentation in the tidally dominated Fly River Delta, Gulf of Papua. *Continental Shelf Research, 13*, 441–472.

Harris, P. T., Tsuji, Y., Marshall, J. F., Davies, P. J., Honda, N., & Matsuda, H. (1996). Sand and rhodolith-gravel entrainment on the mid- to outer-shelf under a western boundary current: Fraser Island continental shelf, eastern Australia. *Marine Geology, 129*, 313–330.

Harris, P. T., Brancolini, G., Armand, L., Busetti, M., Beaman, R. J., Giorgetti, G., Prestie, M., & Trincardi, F. (2001). Continental shelf drift deposit indicates non-steady state Antarctic bottom water production in the Holocene. *Marine Geology, 179*, 1–8.

Harris, P. T., Hughes, M. G., Baker, E. K., Dalrymple, R. W., & Keene, J. B. (2004). Sediment transport in distributary channels and its export to the pro-deltaic environment in a tidally-dominated delta: Fly River, Papua New Guinea. *Continental Shelf Research, 24*, 2431–2454.

Harris, P. T., Heap, A. D., Marshall, J. F., & McCulloch, M. T. (2008). A new coral reef province in the Gulf of Carpentaria, Australia: Colonisation, growth and submergence during the early Holocene. *Marine Geology, 251*, 85–97.

Harris, P. T., MacMillan-Lawler, M., Rupp, J., & Baker, E. K. (2014). Geomorphology of the oceans. *Marine Geology, 352*, 4–24.

Harris, P. T., Alo, B., Bera, A., Bradshaw, M., Coakley, B. J., Grosvik, B. E., Lourenço, N., Moreno, J. R., Shrimpton, M., Simcock, A., & Singh, A. (2016). Chapter 21. Offshore hydrocarbon industries. In L. Inniss, A. Simcock, A. Y. Ajawin, et al. (Eds.), *United Nations World Ocean assessment*. Cambridge: Cambridge University Press.

Haug, G. H., & Tiedemann, R. (1998). Effect of the formation of the isthmus of Panama on Atlantic Ocean thermohaline circulation. *Nature, 393*, 673–676.

Hays, J. D., Imbrie, J., & Shackleton, N. J. (1976). Variations in the Earth's orbit: Pacemaker of the ice ages. *Science, 194*, 1121–1132.

Heezen, B. C., & Hollister, C. D. (1971). *The face of the deep* (p. 659). Oxford University Press, New York.

Heezen, B. C., & Tharp, M. (1965). Tectonic fabric of the Atlantic and Indian oceans and continental drift. *Philosophical transactions of the Royal Society of London. Series A, Mathematical and Physical Sciences, 258*, 90–106.

Heezen, B. C., & Tharp, M. (1977). World ocean floor Panorama, New York. In full color, painted by H. Berann, Mercator Projection, scale 1:23,230,300, 1168 × 1930 mm.

Heinrich, H. (1988). Origin and consequences of cyclic ice rafting in the Northeast Atlantic Ocean during the past 130,000 years. *Quaternary Research, 29*, 142–152.

Hess, H. H. (1962). History of ocean basins. In A. E. J. Engel, H. L. James, & B. F. Leonard (Eds.), *Petrologic studies: A volume in honor of A. F. Buddington* (pp. 599–620). New York: Geological Society of America.

Hill, D. F., Griffiths, S. D., Peltier, W. R., Horton, B. P., & Törnqvist, T. E. (2011). High-resolution numerical modeling of tides in the western Atlantic, Gulf of Mexico, and Caribbean Sea during the Holocene. *Journal of Geophysical Research: Oceans, 116*. https://doi.org/10.1029/2010JC006896.

Hoekstra, A. Y., & Mekonnen, M. M. (2012). The water footprint of humanity. *Proceedings of the National Academy of Sciences, 109*, 3232–3237.

Hollister, C. D., & McCave, I. N. (1984). Sedimentation under deep sea storms. *Nature, 309*, 220–225.

Holpley, D., Smithers, S. G., & Parnell, K. E. (2007). *The geomorphology of the great barrier reef: Development, diversity and change* (p. 546). Cambridge: Cambridge University Press.

Hovland, M. (2008). *Deep-water coral reefs: Unique biodiversity hotspots* (p. 278). Chichester, UK: Praxis Publishing (Springer).

Hughes, T. P., Barnes, M. L., Bellwood, D. R., Cinner, J. E., Cumming, G. S., Jackson, J. B. C., Kleypas, J., van de Leemput, I. A., Lough, J. M., Morrison, T. H., Palumbi, S. R., van Nes, E. H., & Scheffer, M. (2017). Coral reefs in the Anthropocene. *Nature, 546*, 82–90.

Hulbe, C. (2017). Is ice sheet collapse in West Antarctica unstoppable? *Science, 356*, 910–911.

Jackson, L. C., Peterson, K. A., Roberts, C. D., & Wood, R. A. (2016). Recent slowing of Atlantic overturning circulation as a recovery from earlier strengthening. *Nature Geoscience, 9*, 518.

Jakobsson, M., Nilsson, J., Anderson, L., Backman, J., Bjork, G., Cronin, T. M., Kirchner, N., Koshurnikov, A., Mayer, L., Noormets, R., O'Regan, M., Stranne, C., Ananiev, R., Barrientos Macho, N., Cherniykh, D., Coxall, H., Eriksson, B., Floden, T., Gemery, L., Gustafsson, O., Jerram, K., Johansson, C., Khortov, A., Mohammad, R., & Semiletov, I. (2016). Evidence for an ice shelf covering the Central Arctic Ocean during the penultimate glaciation. *Nature Communications, 7*. https://doi.org/10.1038/ncomms10365.

Johnson, G. C. (2008). Quantifying Antarctic bottom water and North Atlantic deep water volumes. *Journal of Geophysical Research: Oceans, 113*. https://doi.org/10.1029/2007JC004477.

Jones, M. R., & Torgersen, T. (1988). Late quaternary evolution of Lake Carpentaria on the Australia - New Guinea continental shelf. *Australian Journal of Earth Science, 35*, 313–324.

Kämpf, J., & Chapman, P. (2016). *Upwelling Systems of the World* (p. 433). Springer, Switzerland.

Kearey, P., Klepeis, K., Vine, F. (2009). Chapter 11, Precambrian tectonics and the supercontinent cycle. In *Global tectonics*, 3rd ed. (pp. 361–377), Wiley-Blackwell, Oxford, UK.

Kidwell, S. M., & Holland, S. M. (2002). The quality of the fossil record: Implications for evolutionary analyses. *Annual Review of Ecology and Systematics, 33*, 561–588.

Koslow, J. A., Auster, P., Bergstad, O. A., Roberts, J. M., Rogers, A., Vecchione, M., Harris, P. T., Rice, J., & Bernal, P. (2015). Biological communities on seamounts and other submarine features potentially threatened by disturbance. In: L. Inniss, A. Simcock, & others (Eds.), *United Nations World Ocean Assessment*. http://www.un.org/Depts/los/global_reporting/WOA_RegProcess.htm

Krausmann, F., Erb, K.-H., Gingrich, S., Haberl, H., Bondeau, A., Gaube, V., Lauk, C., Plutzar, C., & Searchinger, T. D. (2013). Global human appropriation of net primary production doubled in the 20th century. *Proceedings of the National Academy of Sciences, 110*(25), 10324–10329.

Kusahara, K., Hasumi, H., & Williams, G. D. (2011). Impact of the Mertz glacier tongue calving on dense water formation and export. *Nature Communications, 2*, 159. https://doi.org/10.1038/ncomms1156.

Kvenvolden, K. A., & Cooper, C. K. (2003). Natural seepage of crude oil into the marine environment. *Geo-Marine Letters, 23*, 140–146.

Lathe, R. (2004). Fast tidal cycling and the origin of life. *Icarus, 168*, 18–22.

Levitus, S., Antonov, J. I., Boyer, T. P., Locarnini, R. A., Garcia, H. E., & Mishonov, A. V. (2009). Global Ocean heat content 1955–2008 in light of recently revealed instrumentation problems. *Geophysical Research Letters, 36*(7). https://doi.org/10.1029/2008GL037155.

Marsh, L., Huvenne, V. A. I., & Jones, D. O. B. (2018). Geomorphological evidence of large vertebrates interacting with the seafloor at abyssal depths in a region designated for deep-sea mining. *Royal Society Open Science, 5*(8), 180286.

Massom, R., Michael, K., Harris, P. T., & Potter, M. J. (1998). The distribution and formative processes of latent heat polynyas in East Antarctica. *Annals of Glaciology, 27*, 420–426.

Maury, M. F. (1855). *A physical geography of the sea* (310 pp). New York: Harper & Brothers.

Maxwell, W. G. H. (1968). *Atlas of the great barrier reef* (p. 258). Amsterdam: Elsevier.

Mikhail, S., & Sverjensky, D. A. (2014). Nitrogen speciation in upper mantle fluids and the origin of Earth's nitrogen-rich atmosphere. *Nature Geoscience, 7*, 816–819.

Milanković, M. (1941). *Canon of insolation of the earth and its application to the problem of the ice ages* (pp. 1–626). Cemian: Royal Serbian Academy Press.

Miller, J. L. (2017). Ocean currents respond to climate change in unexpected ways. *Physics Today, 70*(1), 17–18.

Muller, R. D., Roest, W. R., Royer, J. Y., Gahagan, L. M., & Sclater, J. G. (1997). Digital Isochrons of the World's ocean floor. *Journal of Geophysical Research, 102*, 3211–3214.

Muller, R. D., Sdrolias, M., Gaina, C., & Roest, W. R. (2008). Age, spreading rates and spreading symmetry of the world's ocean crust. *Geochemistry, Geophysics and Geosystems, 9*(Q04006). https://doi.org/10.1029/2007GC001743

Nagel, T. J., Hoffmann, J. E., & Münker, C. (2012). Generation of Eoarchean tonalite-trondhjemite-granodiorite series from thickened mafic arc crust. *Geology, 40*, 375–378.

Normark, W. R., & Carlson, P. R. (2003). Giant submarine canyons: Is size any clue to their importance in the rock record? *Geological Society of America Special Paper, 370*, 175–190.

Norse, E. A., & Crowder, L. B. (2005). *Marine conservation biology* (p. 470). Washington, D.C.: Island Press.

Nowell, A. R. M., McCave, I. N., & Hollister, C. D. (1985). Contributions of HEBBLE to understanding marine sedimentation. *Marine Geology, 66*, 397–409.

Olson, S. L., Reinhard, C. T., & Lyons, T. W. (2016). Limited role for methane in the mid-Proterozoic greenhouse. *PNAS, 113*(41), 11447–11452.

Overeem, I., Anderson, R. S., Wobus, C. W., Clow, G. D., Urban, F. E., & Matell, N. (2011). Sea ice loss enhances wave action at the Arctic coast. *Geophysical Research Letters, 38*(17). https://doi.org/10.1029/2011GL048681

Pattiaratchi, C. B., & Harris, P. T. (2002). Hydrodynamic and sand transport controls on *en echelon* sandbank formation: An example from Moreton Bay, eastern Australia. *Journal of Marine Research, 53*, 1–13.

Peduzzi, P. (2014). *Sand – Rarer than one thinks. Global environmental alert series* (p. 15). Geneva: UNEP/GRID-Geneva. https://na.unep.net/geas/getuneppagewitharticleidscript.php?article_id=110

Perry, G. D., Duffy, P. B., & Miller, N. L. (1996). An extended data set of river discharges for validation of general circulation models. *Journal of Geophysical Research: Atmospheres, 101*, 21339–21349.

Picard, K., Brooke, B. P., Harris, P. T., Siwabessy, P. J. W., Coffin, M. F., Tran, M., Spinoccia, M., Weales, J., Macmillan-Lawler, M., & Sullivan, J. (2018). Malaysia airlines flight MH370 search data reveal geomorphology and seafloor processes in the remote Southeast Indian Ocean. *Marine Geology, 395*(Supplement C), 301–319.

Piper, D. J. W., Cochonate, P., & Morrison, M. L. (1999). The sequence of events around the epicentre of the 1929 grand banks earthquake: Initiation of debris flows and turbidity current inferred from side scan sonar. *Sedimentology, 46*, 79–97.

Puig, P., Canals, M., Company, J. B., Martı'n, J., Amblas, D., Lastras, G., Palanques, A., & Calafat, A. M. (2012). Ploughing the deep sea floor. *Nature, 489*, 286–289.

Purser, A., Marcon, Y., Hoving, H. J., Vecchione, M., Piatkowski, U., Eason, D., Bluhm, H., & Boetius, A. (2016). Association of deep-sea incirrate octopods with manganese crusts and nodule fields in the Pacific Ocean. *Current Biology, 26*(24), R1268–R1269. https://doi.org/10.1016/j.cub.2016.10.052.

Rose, J. I. (2010). New light on human prehistory in the Arabo-Persian Gulf oasis. *Current Anthropology, 51*, 849–883.

Ryan, W. B. F., & Pitman, W. (1998). *Noah's flood: The new scientific discoveries about the event that changed history*. New York: Simon and Schuster.

Ryan, W. B. F., Major, C. O., Lericolais, G., & Goldstein, S. L. (2003). Catastrophic flooding of the Black Sea. *Annual Review of Earth and Planetary Sciences, 31*, 525–554.

Salinas-de-León, P., Phillips, B., Ebert, D., Shivji, M., Cerutti-Pereyra, F., Ruck, C., Fisher, C. R., & Marsh, L. (2018). Deep-sea hydrothermal vents as natural egg-case incubators at the Galapagos rift. *Scientific Reports, 8*(1), 1788. https://doi.org/10.1038/s41598-018-20046-4.

Schlee, S. (1978). *On almost any wind*. London: Cornell University Press.

Schmandt, B., Jacobsen, S. D., Becker, T. W., Liu, Z., & Dueker, K. G. (2014). Dehydration melting at the top of the lower mantle. *Science, 344*(6189), 1265–1268. https://doi.org/10.1126/science.1253358.

Schmitz, W. J., & McCartney, M. S. (1993). On the North Atlantic circulation. *Reviews of Geophysics, 31*, 29–49.

Schwable, P., Leibmann, B., Köppel, S., & Reiberger, T. (2018) *Assessment of microplastic concentrations in human stool*. Medical University of Vienna, UEG Week Conference, Vienna. http://www.umweltbundesamt.at/fileadmin/site/presse/news_2018/UEG_Week_2018_-_Philipp_Schwabl_Microplastics_Web.pdf

Scotese, C. R. (2014). *Future plate motions & pangea proxima*. Evanston: PALEOMAP Project. http://youtu.be/2It3ETk2MGA

Shedd, J. A. (1928). *Salt from my attic* (p. 63). Mosher Press, Portland, Maine.

Shepard, F. P. (1948). *Submarine geology* (p. 557). New York: Harper & Row.

Smith, C. R., & Baco, A. R. (2003). Ecology of whale falls at the deep sea floor. *Oceanography and Marine Biology: An Annual Review, 41*, 311–354.

Smith, W. H., & Sandwell, D. T. (1997). Global sea floor topography from satellite altimetry and ship depth soundings. *Science Magazine, 277*(5334), 1956–1962.

Som, S. M., Buick, R., Hagadorn, J. W., Blake, T. S., Perreault, J. M., Harnmeijer, J. P., & Catling, D. C. (2016). Earth's air pressure 2.7 billion years ago constrained to less than half of modern levels. *Nature Geoscience, 9*, 448–451. https://doi.org/10.1038/ngeo2713.

Steffen, W., Richardson, K., Rockström, J., Cornell, S. E., Fetzer, I., Bennett, E. M., Biggs, R., Carpenter, S. R., De Vries, W., De Wit, C. A., & Folke, C. (2015). Planetary boundaries: Guiding human development on a changing planet. *Science, 347*(6223), 1259855.

Sumaila, U. R., Lam, V. W. Y., Miller, D. D., Teh, L., Watson, R. A., Zeller, D., Cheung, W. W. L., Côté, I. M., Rogers, A. D., Roberts, C., Sala, E., & Pauly, D. (2015). Winners and losers in a world where the high seas is closed to fishing. *Scientific Reports, 5*, 8481.

Sverdrup, H. U., Johnson, N. W., & Flemming, R. H. (1942). *The oceans* (p. 1087). Prentice Hall: Englewood Cliffs.

Syvitski, J. P. M., Vörösmarty, C. J., Kettner, A. J., & Green, P. (2005). Impact of humans on the flux of terrestrial sediment to the global coastal ocean. *Science, 308*, 376–380.

Tchernia, P. (1980). *Descriptive regional oceanography*. Oxford: Pergamon Marine Series.

Tietsche, S., Notz, D., Jungclaus, J. H., & Marotzke, J. (2011). Recovery mechanisms of Arctic summer sea ice. *Geophysical Research Letters, 38*. https://doi.org/10.1029/2010GL045698.

Turekian, K. K. (2001). Origin of the oceans. In: J. H Steele, S. E. Thorpe, K. K. Turekian (Eds.), *Marine chemistry and geochemistry* (pp. 2055–2058). Academic Press, Cambridge, USA.

Valley, J. W., Cavosie, A. J., Ushikubo, T., Reinhard, D. A., Lawrence, D. F., Larson, D. J., Clifton, P. H., Kelly, T. F., Wilde, S. A., Moser, D. E., & Spicuzza, M. J. (2014). Hadean age for a post-magma-ocean zircon confirmed by atom-probe tomography. *Nature Geoscience, 7*, 219–223.

van den Broeke, M., Box, J., Fettweis, X., Hanna, E., Noël, B., Tedesco, M., van As, D., van de Berg, W. J., & van Kampenhout, L. (2017). Greenland ice sheet surface mass loss: Recent developments in observation and modeling. *Current Climate Change Reports, 3*(4), 345–356.

Van Dover, C. L. (2011). Tighten regulations on deep-sea mining. *Nature, 470*(7332), 31–33.

Van Dover, C. L., Aronson, J., Pendleton, L., Smith, S., Arnaud-Haond, S., Moreno-Mateos, D., Barbier, E., Billett, D., Bowers, K., Danovaro, R., Edwards, A., Kellert, S., Morato, T., Pollard, E., Rogers, A., & Warner, R. (2014). Ecological restoration in the deep sea: Desiderata. *Marine Policy, 44*(0), 98–106.

Van Dover, C. L., Ardron, J. A., Escobar, E., Gianni, M., Gjerde, K. M., Jaeckel, A., Jones, D. O. B., Levin, L. A., Niner, H. J., Pendleton, L., Smith, C. R., Thiele, T., Turner, P. J., Watling, L., & Weaver, P. P. E. (2017). Biodiversity loss from deep-sea mining. *Nature Geoscience, 10*, 464–465.

Varga, P., Rybicki, K. R., & Denis, C. (2006). Comment on the paper "fast tidal cycling and the origin of life" by Richard Lathe. *Icarus, 180*, 274–276.

Verne, J. (1864). *A journey to the center of the earth* (p. 183). Pierre-Jules Hetzel, Paris.

Veron, J. E. N., Hoegh-Guldberg, O., Lenton, T. M., Lough, J. M., Obura, D. O., Pearce-Kelly, P., Sheppard, C. R. C., Spalding, M., Stafford-Smith, M. G., & Rogers, A. D. (2009). The coral reef crisis: The critical importance of <350 ppm CO2. *Marine Pollution Bulletin, 58*, 1428–1436.

Vine, F. J., & Matthews, D. H. (1963). Magnetic anomalies over oceanic ridges. *Nature, 199*, 947–949.

Walker, J. C. G. (1979). The early history of oxygen and ozone in the atmosphere. *Pure Applied Geophysics, 117*, 498–512.

Williams, G. E. (2000). Geological constraints on the Precambrian history of Earth's rotation and the Moon's orbit. *Reviews of Geophysics, 38*(1), 37–60. https://doi.org/10.1029/1999RG900016.

Williams, R., Wright, A. J., Ashe, E., Blight, L. K., Bruintjes, R., Canessa, R., Clark, C. W., Cullis-Suzuki, S., Dakin, D. T., Erbe, C., Hammond, P. S., Merchant, N. D., O'Hara, P. D., Purser, J., Radford, A. N., Simpson, S. D., Thomas, L., & Wale, M. A. (2015). Impacts of anthropogenic noise on marine life: Publication patterns, new discoveries, and future directions in research and management. *Ocean & Coastal Management, 115*(Supplement C), 17–24.

Wilson, J. T. (1966). Did the Atlantic close and then reopen? *Nature, 211*, 676–681.

Young, I. R., Zieger, S., & Babanin, A. V. (2011). Global trends in wind speed and wave height. *Science, 332*(6028), 451–455.

Zhang, Y. (2005). Global tectonic and climatic control of mean elevation of continents, and Phanerozoic sea level change. *Earth and Planetary Science Letters, 237*, 524–531.

Index

© Springer Nature Switzerland AG 2020
P. T. Harris, *Mysterious Ocean*, https://doi.org/10.1007/978-3-030-15632-9